METHODS IN MOLECULAR BIOLOGY

For further volumes:
http://www.springer.com/series/7651

For over 35 years, biological scientists have come to rely on the research protocols and methodologies in the critically acclaimed *Methods in Molecular Biology* series. The series was the first to introduce the step-by-step protocols approach that has become the standard in all biomedical protocol publishing. Each protocol is provided in readily-reproducible step-by-step fashion, opening with an introductory overview, a list of the materials and reagents needed to complete the experiment, and followed by a detailed procedure that is supported with a helpful notes section offering tips and tricks of the trade as well as troubleshooting advice. These hallmark features were introduced by series editor Dr. John Walker and constitute the key ingredient in each and every volume of the *Methods in Molecular Biology* series. Tested and trusted, comprehensive and reliable, all protocols from the series are indexed in PubMed.

Mitotic Exit

Methods and Protocols

Edited by

Victor M. Bolanos-Garcia

Clinical Biochemistry, Oxford Brookes University, Oxford, UK

Humana Press

Editor
Victor M. Bolanos-Garcia
Clinical Biochemistry
Oxford Brookes University
Oxford, UK

ISSN 1064-3745 ISSN 1940-6029 (electronic)
Methods in Molecular Biology
ISBN 978-1-0716-4235-1 ISBN 978-1-0716-4236-8 (eBook)
https://doi.org/10.1007/978-1-0716-4236-8

Cover Caption: The image shows the 3D structure of the Anaphase Promoting Complex/Cyclosome, the master regulator of mitosis exit and a cell in mitosis.

This Humana imprint is published by the registered company Springer Science+Business Media, LLC, part of Springer Nature.
The registered company address is: 1 New York Plaza, New York, NY 10004, U.S.A.

Preface

In eukaryotic organisms mitosis is defined as the process in which a cell divides to give two genetically identical descendant cells. A critical event during mitosis is the detection of errors in chromosome segregation, which is monitored by the spindle assembly checkpoint (SAC), the signalling mechanism that delays mitotic exit until chromosome segregation errors are corrected [1, 2]. SAC signalling is mediated by certain serine/threonine kinases, protein phosphatases, multidomain proteins harbouring diverse functional motifs and the ultimate effector of the SAC, the E3 ubiquitin ligase anaphase-promoting complex/cyclosome (APC/C) [3]. Mitosis exit is tightly regulated in time and space through a complex interplay of protein phosphorylation and dephosphorylation events together with the dynamic association of SAC proteins to form larger assemblies that physically interact with the kinetochore and microtubules [4–11]. Loss of mitotic exit checkpoints is a common feature of cancer cells, leading to mitotic dysregulation, gross chromosome segregation errors and genome instability. Interfering with aberrant chromosome segregation and mitotic exit represents a new widow for the development of more effective therapeutics for the treatment of cancer [12–15].

This book presents a range of state-of-the-art approaches and techniques to study mitotic exit in diverse model organisms. These comprise modern genetic approaches such as transgenic methods generation of RNAi-sensitive cell lines and gene overexpression in heterologue gene expression systems; microscopy techniques such as quantitative live cell imaging and FRET-FLIM for the study of the order of events and of specific protein interactions underpinning mitosis control; biochemical protocols for analysing post-translational modifications responsible of mitotic exit regulation, and methods to promote mitosis arrest in disease-associated conditions such as cancer.

Oxford, UK *Victor M. Bolanos-Garcia*

References

1. McAinsh AD, Kops GJPL (2023) Principles and dynamics of spindle assembly checkpoint signalling. Nat Rev Mol Cell Biol 24, 543–559
2. Lara-Gonzalez P, Pines J, Desai A (2021) Spindle assembly checkpoint activation and silencing at kinetochores. Semin Cell Dev Biol 117, 86–98
3. Peters J-M (2006) The anaphase promoting complex/cyclosome: a machine designed to destroy. Nat Rev Mol Cell Biol 7, 644–656
4. Gama Braga L, Cisneros AF, Mathieu MM, Clerc M, Garcia P, Lottin B, Garand C, Thebault P, Landry CR, Elowe S (2020) BUBR1 pseudokinase domain promotes kinetochore PP2A-B56 recruitment, spindle checkpoint silencing, and chromosome alignment. Cell Rep 33, 108397
5. Broad AJ, DeLuca JG (2020) The right place at the right time: Aurora B kinase localization to centromeres and kinetochores. Essays Biochem 64, 299–311
6. Combes G, Barysz H, Garand C, Gama Braga L, Alharbi I, Thebault P, Murakami L, et al (2018) Mps1 phosphorylates its N-terminal extension to relieve autoinhibition and activate the spindle assembly checkpoint. Curr Biol 28, 872–883

7. Jang JK, Gladstein AC, Das A, Shapiro JG, Sisco ZL, McKim KS (2021) Multiple pools of PP2A regulate spindle assembly, kinetochore attachments and cohesion in Drosophila oocytes. J Cell Sci 134, jcs254037
8. Fischer ES, Yu CWH, Bellini D, McLaughlin SH, Orr CM, Wagner A, Freund SMV, Barford D (2021) Molecular mechanism of Mad1 kinetochore targeting by phosphorylated Bub1. EMBO Rep 22, e52242
9. Sarangapani KK, Koch LB, Nelson CR, Asbury CL, Biggins S (2021) Kinetochore-bound Mps1 regulates kinetochore-microtubule attachments via Ndc80 phosphorylation. J Cell Biol 220, e202106130
10. Kim CH, Kim DE, Kim DH, Min GH, Park JW, Kim YB, Sung CK, Yim H (2022) Mitotic protein kinase-driven crosstalk of machineries for mitosis and metastasis. Exp Mol Med 54, 414–425
11. Manic G, Corradi F, Sistigu A, Siteni S, Vitale I (2017) Molecular Regulation of the Spindle Assembly Checkpoint by Kinases and Phosphatases. Int Rev Cell Mol Biol 328, 105–161
12. Huang HC, Shi J, Orth JD, Mitchison TJ (2009) Evidence that mitotic exit is a better cancer therapeutic target than spindle assembly. Cancer Cell 16, 347–358
13. Sun G, Rong D, Li Z, Sun G, Wu F, Li X, Cao H, Cheng Y, Tang W, Sun Y (2021) Role of small molecule targeted compounds in cancer: progress, opportunities, and challenges. Front Cell Dev Biol 9, 694363
14. Sackton KL, Dimova N, Zeng X, Tian W, Zhang M, Sackton TB, Meaders J, Pfaff KL, Sigoillot F, Yu H, Luo X, King RW (2014) Synergistic blockade of mitotic exit by two chemical inhibitors of the APC/C. Nature 514, 646–649
15. Bolanos-Garcia VM (2009) Assessment of the mitotic spindle assembly checkpoint (SAC) as the target of anticancer therapies. Curr Cancer Drug Targets 9, 131–141

Contents

Contributors

VINCENT ARCHAMBAULT • *Institute for Research in Immunology and Cancer, Université de Montréal, Montréal, QC, Canada; Département de biochimie et médecine moléculaire, Université de Montréal, QC, Canada*

RAÚL BENITO ARENAS • *Departamento de Química Bio-orgánica, IQOG, Madrid, Spain*

AGATHA BASTIDA • *Departamento de Química Bio-orgánica, IQOG, Madrid, Spain*

VICTOR M. BOLANOS-GARCIA • *Faculty of Health and Life Sciences, Department of Biological and Medical Sciences, Oxford Brookes University, Oxford, UK; Oxford Target Therapeutics (OTT), Bioinnovation Hub, Oxford, UK*

ÉRIC BONNEIL • *Institute for Research in Immunology and Cancer, Université de Montréal, Montréal, QC, Canada*

HASSAN BOUSBAA • *UNIPRO - Oral Pathology and Rehabilitation Research Unit, University Institute of Health Sciences (IUCS), Cooperativa de Ensino Superior Politécnico e Universitário (CESPU), Gandra, Portugal*

AYSE KOCA CAYDASI • *Department of Molecular Biology and Genetics, Koç University, Istanbul, Turkey*

MENG-QIU DONG • *National Institute of Biological Sciences, Beijing, China*

VIRGINIE EMOND-FRASER • *Institute for Research in Immunology and Cancer, Université de Montréal, Montréal, QC, Canada; Département de biochimie et médecine moléculaire, Université de Montréal, QC, Canada*

MONIKA ENGELHARDT • *Faculty of Medicine, Department of Medicine I, Medical Center, University of Freiburg, Freiburg, Germany*

HOMA FATMA • *Molecular Cancer Genetics & Translational Research Lab, Section of Genetics, Department of Zoology, Aligarh Muslim University, Aligarh, UP, India*

JULIA FELTHAUS • *Faculty of Medicine, Department of Medicine I, Medical Center, University of Freiburg, Freiburg, Germany*

CHRISTINE GREIL • *Faculty of Medicine, Department of Medicine I, Medical Center, University of Freiburg, Freiburg, Germany*

HAYDEN HOMER • *Christopher Chen Oocyte Biology Research Laboratory, UQ Centre for Clinical Research, Herston, QLD, Australia*

FAN HUANG • *The Key Laboratory of Cell Proliferation and Differentiation of the Ministry of Education, College of Life Sciences, Peking University, Beijing, China*

MARIAM HUDA • *Department of Molecular Biology and Genetics, Koç University, Istanbul, Turkey*

AARON N. JOHNSON • *Department of Developmental Biology, Washington University School of Medicine in St. Louis, St. Louis, MO, USA*

PETER KUBINIOK • *Institute for Research in Immunology and Cancer, Université de Montré al, Montréal, QC, Canada*

MYREILLE LAROUCHE • *Institute for Research in Immunology and Cancer, Université de Montréal, Montréal, QC, Canada*

TINGTING LEI • *The Cancer Institute, The Affiliated Hospital of Qingdao University, Qingdao University, Qingdao, China*

JIE LI • *Beijing Key Laboratory of DNA Damage Response and College of Life Sciences, Capital Normal University, Beijing, China*

Jing Li • *Beijing Key Laboratory of DNA Damage Response and College of Life Sciences, Capital Normal University, Beijing, China*
Miharu Maeda • *Department of Biological Informatics and Experimental Therapeutics, Graduate School of Medicine, Akita University 1-1-1, Hondo, Akita, Japan*
Ammar Mayah • *Oxford Target Therapeutics (OTT), Bioinnovation Hub, Oxford, UK*
Daniel Moreno-Andrés • *Institute of Biochemistry and Molecular Cell Biology, Medical School, RWTH Aachen University, Aachen, Germany*
Charlotte Pain • *Endomembrane Structure and Function Research Group, Department of Biological and Medical Sciences, Oxford Brookes University, Oxford, UK*
Cristina Mirela Pascariu • *Institute for Research in Immunology and Cancer, Université de Montréal, Montréal, QC, Canada*
James R. Paulson • *Department of Chemistry, University of Wisconsin-Oshkosh, Oshkosh, WI, USA*
Bin Peng • *Guangdong Key Laboratory for Genome Stability & Disease Prevention and Carson International Cancer Center, Marshall Laboratory of Biomedical Engineering, Shenzhen University School of Medicine, Shenzhen, Guangdong, China*
Bárbara Pinto • *UNIPRO - Oral Pathology and Rehabilitation Research Unit, University Institute of Health Sciences (IUCS), Cooperativa de Ensino Superior Politécnico e Universitário (CESPU), Gandra, Portugal*
Kota Saito • *Department of Biological Informatics and Experimental Therapeutics, Graduate School of Medicine, Akita University 1-1-1, Hondo, Akita, Japan*
Anja Scheufen • *Institute of Biochemistry and Molecular Cell Biology, Medical School, RWTH Aachen University, Aachen, Germany*
Guangcan Shao • *National Institute of Biological Sciences, Beijing, China*
Hifzur R. Siddique • *Molecular Cancer Genetics & Translational Research Lab, Section of Genetics, Department of Zoology, Aligarh Muslim University, Aligarh, UP, India*
João P. N. Silva • *UNIPRO - Oral Pathology and Rehabilitation Research Unit, University Institute of Health Sciences (IUCS), Cooperativa de Ensino Superior Politécnico e Universitário (CESPU), Gandra, Portugal*
Patrícia M. A. Silva • *UNIPRO - Oral Pathology and Rehabilitation Research Unit, University Institute of Health Sciences (IUCS), Cooperativa de Ensino Superior Politécnico e Universitário (CESPU), Gandra, Portugal; Associate Laboratory i4HB - Institute for Health and Bioeconomy, UCIBIO - Applied Molecular Biosciences Unit, Translational Toxicology Research Laboratory (1H-TOXRUN, IUCS-CESPU), University Institute of Health Sciences, CESPU, CRL, Gandra, Portugal*
Pierre Thibault • *Institute for Research in Immunology and Cancer, Université de Montré al, Montréal, QC, Canada; Département de chimie, Université de Montréal, Montréal, QC, Canada*
Ralph Wäsch • *Faculty of Medicine, Department of Medicine I, Medical Center, University of Freiburg, Freiburg, Germany*
Xingzhi Xu • *Guangdong Key Laboratory for Genome Stability & Disease Prevention and Carson International Cancer Center, Marshall Laboratory of Biomedical Engineering, Shenzhen University School of Medicine, Shenzhen, Guangdong, China*
Shige Yamada • *Department of Developmental Biology, Washington University School of Medicine in St. Louis, St. Louis, MO, USA*
Shuo Yang • *Department of Genetics and Genetics Engineering, School of Life Science, Fudan University, Shanghai, China*

YUNAN YE • *Christopher Chen Oocyte Biology Research Laboratory, UQ Centre for Clinical Research, Herston, QLD, Australia*
CHUANMAO ZHANG • *The Academy for Cell and Life Health, Faculty of Life Science and Technology, Kunming University of Science and Technology, Kunming, China; The Key Laboratory of Cell Proliferation and Differentiation of the Ministry of Education, College of Life Sciences, Peking University, Beijing, China*
GANG ZHANG • *The Cancer Institute, The Affiliated Hospital of Qingdao University, Qingdao University, Qingdao, China*
CHENXI ZHOU • *Christopher Chen Oocyte Biology Research Laboratory, UQ Centre for Clinical Research, Herston, QLD, Australia*

Chapter 1

A Transgenic Method to Measure Mitotic Exit in *Drosophila* Embryos

Shige Yamada, Aaron N. Johnson, and Shuo Yang

Abstract

Mitotic exit is a necessary step for highly specialized cells to terminally differentiate and acquire unique functions. The FUCCI system can be used to visualize mitotic and post-mitotic cells during development and regeneration in both live organisms and fixed tissues. Here we describe a Fly-FUCCI protocol for assaying mitotic exit in *Drosophila* embryos.

Key words Mitotic exit, Fly-FUCCI, Myogenesis, *Drosophila* embryogenesis, Mef2

1 Introduction

Cell cycle exit is an essential step of differentiation. Highly specialized cells, such as striated muscles, assemble complex cytoplasmic structures that are incompatible with cell proliferation. For example, embryonic heart cells (cardiomyocytes) in mice contract during development to establish blood flow in the embryo, but cardiomyocytes also proliferate as the embryonic heart grows. Cardiomyocyte proliferation requires the cells to disassemble the contractile machinery (sarcomeres) before reentering the cell cycle [1]. In salamanders, syncytial skeletal muscle cells (myofibers) fragment in response to injury, dedifferentiate, and reenter the cell cycle. The dedifferentiated cells then proliferate, differentiate back into muscle precursors (myoblasts), and repair the damaged tissue [2]. Visualizing mitotic exit in vivo is therefore necessary to understand cellular differentiation during development and regeneration.

Replication proteins (Ki-67, PCNA) and post-transcriptional modifications (phosphorylation) to Histone 3 (H3) have traditionally been used to identify mitotic cells [3, 4]. Antibodies are commercially available that recognize mitotic markers, but the abundance and localization of each antigen varies across the cell cycle. Fluorescent Ubiquitous Cell Cycle Indicator (FUCCI)

Victor M. Bolanos-Garcia (ed.), *Mitotic Exit: Methods and Protocols*, Methods in Molecular Biology, vol. 2874,
https://doi.org/10.1007/978-1-0716-4236-8_1,

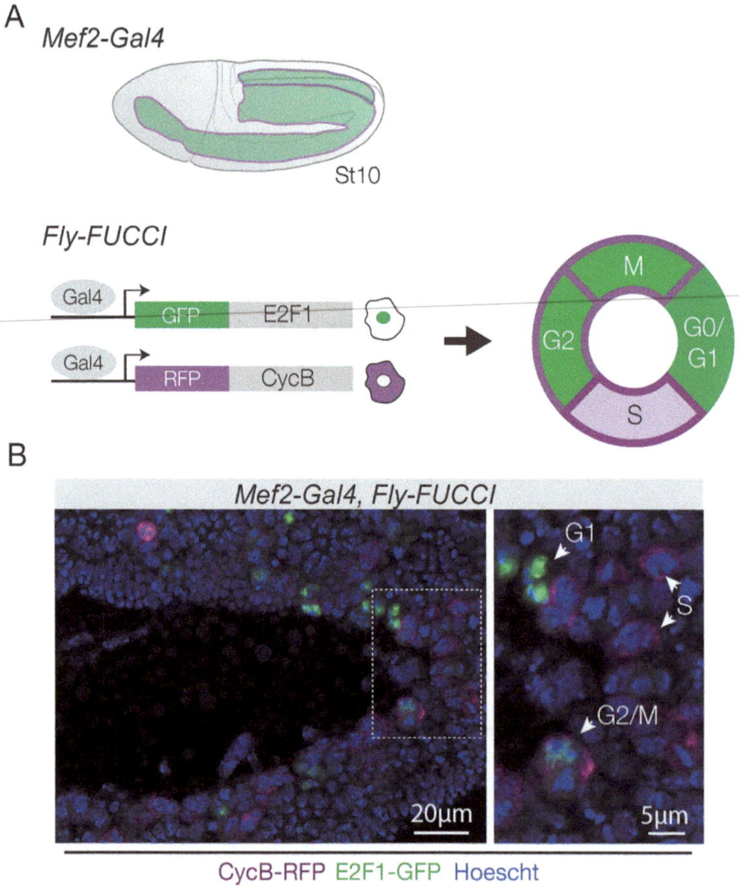

Fig. 1 Fly-FUCCI system. (**a**). Diagram of the *Mef2-Gal4* expression in Stage 10 (St 10) *Drosophila* embryos and the cell cycle–specific expression of the FUCCI proteins. The FUCCI system uses chimeric proteins that contain fluorescence domains (GFP or dsRED) and cell cycle proteins (E2F1 and CycB). CycB is targeted for degradation during G0 and G1; E2F1 is targeted for degradation during S-phase. Each stage of the cell cycle shows a distinct fluorescence pattern. (**b**) Confocal micrograph of *Mef2-Gal4* driving Fly-FUCCI in the embryonic mesoderm at St10. Mitotic cells express both GFP and dsRed

technologies use transgenic fluorescent proteins to unambiguously identify each stage of the cell cycle in live organisms and fixed tissues [5]. In addition, the FUCCI system can be expressed with cell type specificity to identify cells within a particular lineage that have exited the cell cycle. We used Fly-FUCCI (Fig. 1) to track cell cycle progression in muscle precursors of *Drosophila* embryos and found expression of Aurora kinases alone is sufficient to inhibit mitotic exit [6]. The FUCCI system is thus a robust tool to visualize mitotic exit in vivo during embryogenesis.

2 Materials

2.1 *Drosophila* Stocks

1. *Mef2-Gal4* (Bloomington Stock Center 27390), for muscle-specific expression.
2. Fly-FUCCI. *UAS-GFP.E2F1*$_{1\text{–}230}$, *UAS-mRFP1.CycB*$_{1\text{–}266}$ (Bloomington Stock Center 55117). The mRFP1 transgene is a derivative of the dsRED locus in *Discosoma*, which is distinct from the GFP locus in *Aequorea*. The GFP.E2F1 and mRFP1.CycB are detected with distinct antibodies. The Fly-FUCCI is described in detail in [5] and shown in Fig. 1.

2.2 Antibodies

1. α-GFP (1:300, Aves Labs, GFP-1020), α-chicken HRP (1:150, Aves Labs, IRH-1010).
2. α-dsRed (1:300, Takara, 632,392), α-rabbit HRP (included in Tyramide Signal Amplification Kit, Thermo B40922).
3. α-βgal (1:100, Promega, Z3781), α-mouse HRP (Jackson ImmunoResearch Laboratories, 115-585-003).
4. Tyramide Signal Amplification Kits (Thermo B40922, B40942) to detect GFP and dsRed.

2.3 Equipment

1. Embryo collection cages with 6 cm grape plates.
2. Grape plate: 0.4% agarose (40 g), grape juice (500 mL). Adjust volume to 1 L with milliQ water. Heat in microwave until the agarose completely dissolved. Cool to 50 °C, add 0.5 mL ethanol, 10 mL acetic acid, mix and pour into 6 cm plates. Store at 4 °C for up to 1 month.
3. 100 μm Cell strainers.
4. Glass scintillation vials, 7 mL size.
5. 6-well tissue culture plate.
6. 9-well glass plate.

2.4 Reagents and Chemicals

1. Prepare solutions with ultrapure sterilized water and store at room temperature unless otherwise indicated.
2. 10× PBS: 1.37 M NaCl (80 g), 27 mM KCl (2 g), 81 mM Na_2HPO_4 (11.5 g) 15 mM KH_2PO_4 (2 g). Adjust pH to 7.4 with 10 N NaOH. Adjust volume to 1 L. Sterilize by autoclaving.
3. PBT: 100 mL 10× PBS, 1 ml Tween 20 per 1 L.
4. PBST: 100 mL 10× PBS, 1 ml Triton ×-100 per 1 L.
5. PEM: 0.1 M PIPES (30.2 g), 2 mM $MgS0_4$ (2 ml of 1 M stock), 1 mM EGTA (0.38 g) per 1 L.

6. 25%NGS, PBST 5%: 2.5 mL Goat Serum, 5 mL 10% Triton ×-100, 1 mL 10× PBS. Adjust volume to 10 mL with water. Store at 4 °C for up to 1 week.
7. 5%NGS/PBT: 0.5 mL Goat Serum, 1 mL 10XPBS, 10 μL Tween 20 per 10 mL. Store at 4 °C for up to 1 week.
8. 16% formaldehyde (methanol-free, Thermo Fisher 28908). Open by breaking the glass ampule. Transfer to solution to a 15 mL tube and store at −20 °C.
9. 30% (w/w) H_2O_2.
10. Heptane.
11. Household bleach.

3 Method

3.1 Embryo Collection, Dechlorination, and Fixation

1. Prepare yeast mixture for embryo collection. Put 15 mL dry yeast in a 50 mL centrifuge tube, add 15 mL sterile water, and mix well. A good mixture should be wet enough to stick to a grape plate but dry enough for flies to walk on.
2. Warm a grape plate to 25 °C.
3. Spread yeast mixture in the center of a grape plate; the final diameter of the mixture should be around 2–3 cm.
4. Collect ~50 newly eclosed (within 3 ~ 5 days) *Mef2-Gal4* virgin flies and 10 *UAS-GFP.E2F1*$_{1-230}$, *UAS-mRFP1.CycB*$_{1-266}$ male flies. Place male and female flies in an embryo collection cage, add the grape plate with yeast, and incubate the flies at 25 °C with 60% humidity under a 12 h light/dark cycle.
5. Change the grape plate every 24 h which will yield a mixture of embryos at all developmental stages. Put the grape plate with embryos to 4 °C to stop embryonic development. The embryos can be stored at 4 °C for 2 days. Since female fecundity decreases with time, embryos should not be collected for more than 7 days.
6. Wash the plate with deionized (DI) water and use paint brush to dissolve the yeast and free embryos from agar. Move the embryos to a 100 μm cell strainer, and rinse embryos with DI water to remove all the yeast. Healthy embryos will be white.
7. Remove embryo chorions. Place the cell strainer on a laboratory wipe to remove excess water, then transfer to a 50% bleach solution in a 6-well plate for 2:30 min with gentle shaking. Immediately remove the bleach solution by wipes and rinse the embryos extensively with DI water for few times. Check the dechlorination status under a stereoscopic microscope. 50%

bleach solution will lose activity after 1 week, making chorion removal inefficient.

8. Embryos can be live imaged at this point to visualize cell cycle dynamics. Our live imaging protocol is described in [7].
9. To continue with immunohistochemistry, transfer the embryos from the cell strainer to a scintillation vial with PBST, wash the embryos with PBST three times, then remove all PBST.
10. Add 1.5 mL PEM solution, 2 mL heptane, and 0.5 mL 16% formaldehyde to the scintillation vial. Two layers will be present: The lower layer is fixative and the upper layer is heptane. Embryos will settle at the interface between the layers. If the embryos sink to the bottom of the vial remake or replace the solutions.
11. Shake the scintillation vial on a laboratory shaker at 200 rpm for 15 min. The heptane will become saturated with fixative, allowing the fixative to penetrate the hydrophobic vitelline membrane surrounding the embryos and ensure the fixation.
12. Wait 3–5 min until the interface completely settles. There should still be two phases in the vial, with a majority of the embryos at the interface. Remove the lower layer from the scintillation vial along with any embryos that fell from the interface.
13. Remove vitellin membranes. Add 2 mL methanol to the scintillation vial and shake by hand for ~15 s. Embryos will sink to the bottom of the vial when the vitellin membranes rupture. Allow the embryos to settle for 3–5 min. Collect the embryos at the bottom of the vial and transfer them into a 1.5 mL tube. The empty vitelline membranes and non-devitellinized embryos will remain at the interface between the heptane layer and methanol layer.
14. Wash the embryos with methanol three times to remove residual heptane. Embryos can be stored in −20 °C for up to 2 years.

3.2 Embryo Staining

Day 1

1. Remove the embryos from −20 °C and wash with methanol three times. All washes are 1 mL unless otherwise noted.
2. Add 0.9 mL methanol and 0.1 mL 30% H_2O_2, rotate at room temperature for 2 h (excessive incubation might damage the embryos). This step quenches endogenous peroxidase and can significantly reduce staining background when using Tyramide Signal Amplification.

3. Wash the embryos with methanol three times. Add 0.5 mL methanol and 0.5 mL PBT, rotate at room temperature for 5 min.
4. Remove the liquid and wash once with PBT, then incubate with PBT for 5 min while rotating.
5. Wash the embryos with PBT once more, then add 1 mL blocking solution (25%NGS/PBST 5%), and rotate end over end at room temperature for 1 h.
6. Remove the liquid and wash once with 5%NGS/PBT. Dilute the primary antibodies (anti-GFP and anti-dsRed) in 300 μL staining solution and incubate end over end at 4 °C overnight.

Day 2

7. Remove the primary antibody and wash the embryos with staining solution four times for 10 min each.

   Note: DO NOT discard the primary antibody, normally the primary antibody can be reused at least four times. The reused antibody often yields better staining results with lower background.
8. Add 2 μL goat anti-Chicken HRP secondary antibody to 300 μl PBT, rotate end over end for 2 h at room temperature.
9. Remove the secondary antibody and wash the embryos with PBT for 4× for 10 min each.
10. Protect the embryos from light from this step. Make TSA reaction buffer per manufacturer's instructions. Add 200 μL reaction buffer with AF555 Tyramide and rotate for 4–7 min. Longer incubation time can result in stronger staining but also increase the background. For dsRed, 4 min is often suitable.
11. Remove the reaction buffer but do not discard. Wash the embryos with PBT (three quick washes and one 10 min rotating wash) and check the staining result under a dissecting fluorescent microscope. **Steps 10–11** can be repeated with saved reaction buffer until fluorescence signal is optimal.
12. Quench the α-chicken HRP. Add 0.9 ml PBT and 0.1 mL 30% H_2O_2, incubate at 4 °C overnight gently rotating.

Day 3

13. Remove the PBT/H_2O_2, wash the embryos with PBT (three quick washes, one 1 h rotating wash, one quick wash).
14. Add 300 μL α-rabbit HRP from the TSA kit, rotate gently for 2 h at room temperature.
15. Repeat **steps 9–11**, substituting Tyramide AF488 for AF555. For GFP, 4 min TSA reaction is often suitable.

16. Remove the reaction buffer but do not discard. Wash the embryos with PBT (three quick washes and one 10 min rotating wash) and check the staining result under a dissecting fluorescent microscope. **Step 15** can be repeated with saved reaction buffer until fluorescence signal is optimal.
17. Add Hoechst solution (1:1000 in PBT) and incubate for 2 min while rotating.
18. Wash the embryos with PBT (three quick washes and one 10 min rotating wash).
19. Remove the buffer and add 1 mL mounting medium (70% glycol/30% PBT), the embryos should be stored at 4 °C and protected from light.

3.3 Mount the Embryo

1. Transfer embryos to one well of a 9-well glass plate. Use a dissecting microscope to identify an embryo at the desired developmental stage. Transfer the embryo in ~50 μL mounting medium to a clean slide.
2. Spray a small amount of plasticine on each of the four corners of 22 mm square coverslip to make a raised coverslip.
3. Place the coverslip on the slide. Gently press the four corners of the coverslip. Check the orientation of the embryo under the microscope, gently push the edge of coverslip vertically or horizontally to optimize the embryo position.

3.4 Imaging

1. Use a confocal system to image the embryos. Normally, 200× magnification lens is used to image the entire embryo, and 400× magnification is used for higher resolution. The laser intensity and master gain can be modified based on the staining result.
2. Set the scanning interval to 1 μm. 30–40 scans are usually needed to image the body wall muscles.

3.5 Interpretation and Quantification

1. Fly-FUCCI will express GFP in G0 and G1 cells, dsRed in S-phase cells, and GFP/dsRed in G2/M cells (Fig. 1).
2. ImageJ can be used to count the number of each cell type in an image stack.

4 Notes

1. The *Mef2-Gal4* stock directs expression of the Fly-FUCCI transgenes in a cell type of interest. *Mef2-Gal4* is active throughout the embryonic muscle lineage. Gal4 stocks are publicly available that are active in a myriad of cell types.

2. The incubation time of the 50% bleach can be extended to achieve complete dechlorination.
3. Generally we fix the embryos for 15 min, but the time can be adjusted based on the antigens and antibodies.
4. Plasticine on coverslip corners is used to provide a safe space between the coverslip and the embryo. To make the imaging chamber, never press the corners too hard, which will crush the embryo. Plasticine from different suppliers may have different physical properties, the soft and sticky plasticine is preferred.
5. After staining, slides can be stored at 4 °C for up to 6 months.
6. For confocal imaging, we use the "range indicator" function to minimize pixel saturation.

References

1. Ahuja P, Perriard E, Perriard JC, Ehler E (2004) Sequential myofibrillar breakdown accompanies mitotic division of mammalian cardiomyocytes. J Cell Sci 117:3295–3306. https://doi.org/10.1242/jcs.01159
2. Wang H, Simon A (2016) Skeletal muscle dedifferentiation during salamander limb regeneration. Curr Opin Genet Dev 40:108–112. https://doi.org/10.1016/j.gde.2016.06.013
3. von Bohlen und Halbach, O. (2011) Immunohistological markers for proliferative events, gliogenesis, and neurogenesis within the adult hippocampus. Cell Tissue Res 345:1–19. https://doi.org/10.1007/s00441-011-1196-4
4. Muskhelishvili L, Latendresse JR, Kodell RL, Henderson EB (2003) Evaluation of cell proliferation in rat tissues with BrdU, PCNA, Ki-67 (MIB-5) immunohistochemistry and in situ hybridization for histone mRNA. J Histochem Cytochem 51:1681–1688. https://doi.org/10.1177/002215540305101212
5. Zielke N, Edgar BA (2015) FUCCI sensors: powerful new tools for analysis of cell proliferation. Wiley Interdiscip Rev Dev Biol 4:469–487. https://doi.org/10.1002/wdev.189
6. Yang S, McAdow J, Du Y, Trigg J, Taghert PH, Johnson AN (2022) Spatiotemporal expression of regulatory kinases directs the transition from mitosis to cellular morphogenesis in drosophila. Nat Commun 13:772. https://doi.org/10.1038/s41467-022-28322-8
7. Yang S, Weske A, Du Y, Valera JM, Jones KL, Johnson AN (2020) FGF signaling directs myotube guidance by regulating Rac activity. Development 147. https://doi.org/10.1242/dev.183624

Chapter 2

Generation of Cdc20 RNAi-Sensitive Cell Lines to Study Mitotic Exit

Tingting Lei and Gang Zhang

Abstract

Accurate mitotic progression ensures the fidelity of genome passage. Cdc20 is a key mitotic regulator. It promotes mitotic exit by activating the anaphase-promoting complex or cyclosome (APC/C) and monitors kinetochore-microtubule attachment through activating the spindle assembly checkpoint (SAC). Precise characterization of Cdc20 requires efficient depletion of endogenous Cdc20, which is extremely difficult to achieve by RNA interference (RNAi). This chapter describes the methodology to generate Cdc20 RNAi-sensitive cell lines with the help of CRISPR/Cas9 technology. These cell lines are highly sensitive to Cdc20 RNAi and provide a very useful tool for Cdc20 functionality investigation without the interference of endogenous Cdc20 protein. Similar strategy could be applied to other genes.

Key words Cdc20, CRISPR/Cas9, RNAi

1 Introduction

Cdc20, an essential mitotic regulator, plays opposite roles in mitosis to ensure the fidelity of chromosome segregation. On one hand, Cdc20 directly activates APC/C by binding and promoting conformational changes of APC/C. Cdc20 also recognizes degrons on APC/C substrates and facilitates their ubiquitination. On the other hand, unattached kinetochores activate SAC resulting in the production of MCC complex composed by Cdc20 and three other checkpoint proteins. MCC binds and inhibits the activity of APC/C^{Cdc20}. Only when all kinetochores have been attached by microtubules, SAC is turned off and MCC is disassembled. APC/C^{Cdc20} regains the activity to ubiquitinate cyclin B1 and securin. Anaphase is initiated with declining Cdk1 activity and separating chromatids. The interplay between MCC and APC/C determines the timely mitotic exit [1].

Cdc20 has been extensively studied. One challenge of studying Cdc20 is that the endogenous Cdc20 is very difficult to be

Victor M. Bolanos-Garcia (ed.), *Mitotic Exit: Methods and Protocols*, Methods in Molecular Biology, vol. 2874,
https://doi.org/10.1007/978-1-0716-4236-8_2,

efficiently depleted by RNAi. It is estimated that endogenous Cdc20 has to be lower than 5% to arrest cells at metaphase [2, 3]. Thus, residual Cdc20 may interfere the mitotic progression in experiments exploring exogenous Cdc20 functionality. Other methods like knock in inducible degron tag also face the same challenges by the residual protein besides the technical difficulty of low knock in efficiency. Physical removal of Cdc20 exons by gene trap method causes apoptosis [4]. In theory, knock out Cdc20 by CRISPR/Cas9 technology will also lead to cell death. However, in the last few years, mounting evidences show genes being knocked out by CRISPR/Cas9 may not be fully silenced [5]. Through alternative splicing or translation reinitiation, residual protein could be produced in knockout cells/organisms maintaining partial or even full functionality which has been termed as knockout escaping. A rough estimation indicates around 30–50% of the knockout cells/organisms are producing altered transcripts and proteins. In some cases, the residual protein is detectable, while in other cases it is not because of the low expression level, the lack of suitable antibody, the changes on the antigens and so on. Because of the low expression level of the functioning residual protein, RNAi-sensitive cell lines could be generated and maintained in labs, especially for the essential genes like *CDC20* whose protein product is only needed at minimal levels.

Based on this observation, we have successfully generated RNAi-sensitive cell lines for three checkpoint genes *CDC20, ROD,* and *BUB1* with the help of CRISPR/Cas9 technology (Fig. 1). In all cases, RNAi treatment induces strong mitotic defects which could not be achieved in wild type cells [6, 7].

2 Materials

2.1 Reagents

BbsI.

Cdc20 antibody.

GAPDH antibody.

Genomic DNA purification kit.

Fluorophore-labeled secondary antibody.

Lipofectamine 2000.

Lipofectamine RNAiMax.

Nocodazole.

pcDNA5/FRT/TO N-YFP tag expression plasmid.

PCR purification kit.

Plasmid miniprep kit.

Protein A/G PLUS-Agarose.

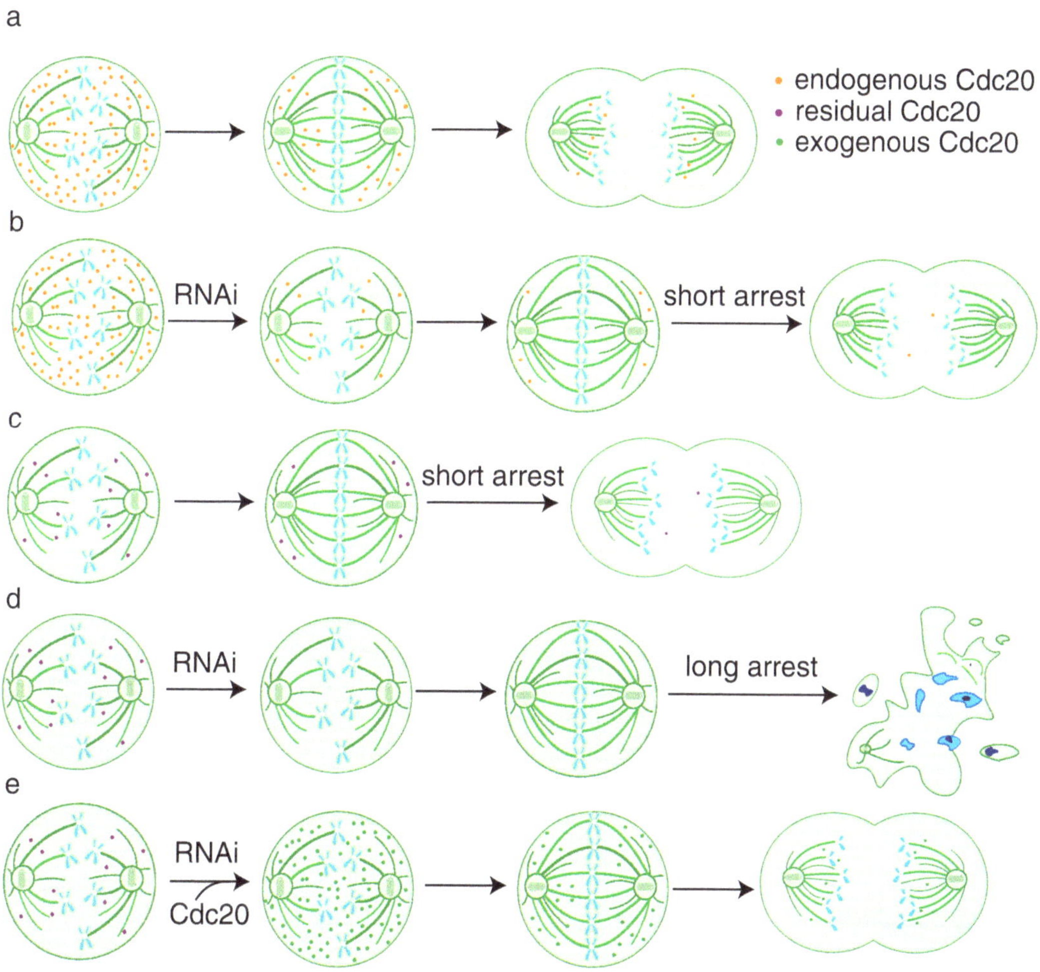

Fig. 1 Cartoon showing mitotic progression of HeLa and Cdc20 KO cells under different circumstances. (**a**) HeLa cells show normal mitotic progression. (**b**) Knockdown Cdc20 by RNAi largely reduces endogenous Cdc20 in HeLa cells but only causes short metaphase arrest before mitotic exit. (**c**) Cdc20 knockout cells express residual Cdc20 and show delayed mitosis with short metaphase arrest. (**d**) Completely depleting Cdc20 by RNAi in Cdc20 knockout cells fully blocks mitotic progression by arresting cells at metaphase till apoptosis. (**e**) Introducing exogenous Cdc20 to (**d**) fully restores normal mitotic progression

Puromycin.

px459.

T4 DNA ligase.

TA clone kit.

Thymidine.

2.2 Cell Lines and Medium

HeLa cell line.

Dulbecco's modified Eagle medium (DMEM).

Fetal bovine serum (FBS).

0.05% trypsin-EDTA.

1% penicillin-streptomycin (PenStrep).

Sterile phosphate-buffered saline (PBS).

Improved Minimal Essential Medium (Opti-MEM).

Leibovitz's L-15 medium.

2.3 General Equipment

CO_2 incubator.

Confocal or fluorescence microscopy.

Stage top incubator.

Thermocycler.

Western blot imaging system.

3 Methods

The methodology described here includes the designing of sgRNA targeting *CDC20* and cloning of plasmid expressing Cas9 and sgRNA; the generation, screening, and confirmation of Cdc20 knockout cell lines; RNAi treatment of Cdc20 knockout cells and rescue of the metaphase arrest by introducing exogenous Cdc20. Two figures (Figs. 2, 3) showing the absence of Cdc20 in knockout cells by Western blot and exogenous wild type Cdc20, but not Cdc20ΔIR could promote mitotic exit by live cell imaging are presented as examples of validity of the method.

3.1 Cdc20 sgRNA Design

1. Generate an account in Benchling (www.benchling.com) and log in.
2. Create a project for CRISPR guides. Type in *CDC20* as the target gene and choose hg38 (*Homo sapiens*) as the reference genome. Choose single guide in the design type window.
3. Choose exon 2 by clicking and dragging the mouse. The position of chosen genome region appears in the right half window. Press "+" to show all the possible guide sequences within the chosen region. Choose the guides with high on-target score and off-target score. For example, we chose the one with on-target score 77.4 and off-target score 41.3 (5′ ACCACTCCTAGCAAACCTGG 3′) in our study (*see* **Note 1**).

3.2 CDC20 Knockout Construct Cloning

1. Design and synthesize cloning primers for sgRNA cloning (*see* **Note 2**).

   Forward primer: 5′ CACCGACCACTCCTAG CAAACCTGG 3′.

   Reverse primer: 5′ AAACCCAGGTTTGCTAG GAGTGGTC 3′.

2. Dilute both primers to 100 μM. Mix 10 μL of each primer and heat at 95 °C for 8 min. Turn off the heating block to let the mixture cool down to room temperature.
3. Digest 1.5 μg of px459 vector with 1.0 μL of restrictive enzyme BbsI in a total volume of 50 μL at 37 °C for 1 h. Purify the digested vector using PCR purification kit according to the instructions.
4. Mix 7.5 μL of primer mixture with 1.5 μL of purified px459, 1 μL of T4 DNA ligase buffer, and 0.2 μL of T4 DNA ligase. Incubate at room temperature for 3 h.
5. Transform 1 μL of ligate into 50 μL of competent cells (*E. coli* DH5α). Streak bacterial onto LB plate with 100 μg/mL ampicillin and incubate at 37 °C overnight.
6. Pick up and cultivate a few colonies. Isolate plasmids with plasmid miniprep kit and send for sequencing by primer: 5′ TTTATGGCGAGGCGGCGG 3′.

3.3 Cdc20 Knockout Cell Lines Generation

1. Seed HeLa cells in 6-well plate at 25% confluency with DMEM medium (supplemented with 10% FBS and PenStrep). Cultivate cells at 37 °C in CO_2 incubator for 24 h.
2. Dilute 500 ng of px459 plasmid with cloned sgRNA in 0.25 mL of Opti-MEM. Dilute 5 μL of Lipofectamine 2000 in 0.25 mL of Opti-MEM. Incubate at room temperature for 5 min. Transfer diluted Lipofectamine 2000 into diluted DNA and mix well by pipetting gently several times. Incubate the mixture at room temperature for 20 min.
3. Wash HeLa cells once with 2 mL of PBS and add 2 mL of Opti-MEM in the well during the incubation. Add DNA/Lipofectamine 2000 complex (0.5 mL) dropwise into the well with HeLa cells. After 6 h, replace Opti-MEM with fresh DMEM medium. Incutate the cells at 37 °C in CO_2 incubator for 24 h.
4. Collect all cells by trypsinization and reseed in 15 cm dish with freshly prepared DMEM medium containing 1 μg/mL puromycin. Incubate cells at 37 °C in CO_2 incubator for 48 h.
5. Remove the selection medium and replace with fresh DMEM medium. Keep cultivating cells till colonies grow up which takes two to three weeks.
6. Pick up individual colony with sterile pipette tip and cultivate in 24-well plate. Amplify cells for further characterization.

3.4 Cdc20 Knockout Screening

1. Seed parental HeLa cells and the cells isolated from 3.3 into 6-well plate at 50% of confluency. Add thymidine (2 mM) to the DMEM medium and incubate cells at 37 °C in CO_2 incubator for 24 h.

2. Wash cells once with 2 mL of PBS and continue cultivation with fresh DMEM medium at 37 °C in CO_2 incubator for 24 h.
3. Change medium with DMEM containing 2 mM of thymidine. Cultivate cells at 37 °C in CO_2 incubator overnight.
4. Wash cells once with 2 mL of PBS and add fresh DMEM medium. Keep cultivating at 37 °C in CO_2 incubator for 8 h.
5. Collect all cells by trypsinization and wash cells once with PBS.
6. Resuspend cells in lysis buffer (10 mM Tris/Cl pH 7.5, 150 mM NaCl, 0.5 mM EDTA, 0.5% NP40 with protease inhibitor). Leave cells on ice for 30 min before centrifuge at 17,000× *g* for 10 min at 4 °C.
7. Transfer the supernant to 1.5 mL Eppendorf tubes. Measure protein concentration.
8. Take 30 μg of cell lysate of each sample, mix with same volume of 2 × SDS-sample buffer. Boil at 95 °C for 5 min.
9. Load each sample side by side onto SDS-PAGE gel and separate protein by electrophoresis.
10. Transfer separated protein to PVDF membrane and block the membrane with 5% nonfat milk/PBST (PBS with 0.1% Tween-20) for 1 h.
11. Incubate the membrane with 5% nonfat milk/PBST with Cdc20 antibody (1:1000) and GAPDH antibody (1:1000) at room temperature for 1 h (*see* **Note 3**).
12. Wash the membrane three times with PBST and incubate with fluorophore-labeled secondary antibody (1:5000) in 5% nonfat milk/PBST for 30 min.
13. Detect the presence of Cdc20 and GAPDH with Western blot imaging system (Fig. 2).
14. Amplify cells with no detectable Cdc20 and freeze at −80 °C.

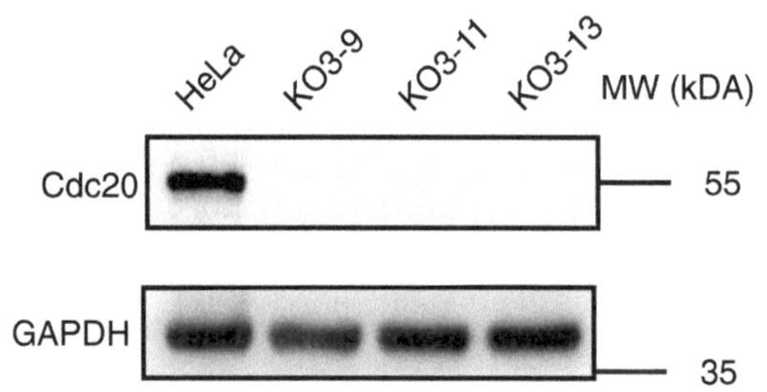

Fig. 2 Western blot showing the absence of endogenous Cdc20 in Cdc20 knockout cells. Cell lysate from parental HeLa cells and three Cdc20 knockout cell lines are examined for the existence of Cdc20 by Western blot with antibody against the C-terminal region of Cdc20. Cdc20 is only detected in HeLa cells, but not in the three knockout cell lines

3.5 Cdc20 Knockout Confirmation by Genomic PCR

1. Collect 1×10^6 of Cdc20 knockout cells and parental HeLa cells as control. Purify genomic DNA with genomic DNA purification kit according to the instructions.
2. Synthesize the following primers (*see* **Note 4**).
   Forward: 5′ CAAAGCCAAGGAAGC 3′.
   Reverse: 5′ GTTCAGGTTCAAAGC 3′.
3. Take 200 ng of genome DNA as template and amplify DNA using the above primers and Taq DNA polymerase (*see* **Note 5**).
4. Set up TA cloning reaction according to kit instructions.
5. Transform *E. coli* DH5α with the TA reaction solution.
6. Pick up six colonies. Purify plasmids and send for sequencing by primer suitable for the TA vector.
7. Analyze the sequence to confirm the presence of indels generated by CRISPR/Cas9 and non-homologous end-joining (NHEJ).

3.6 Identification of Residual Cdc20 in Cdc20 Knockout Cells by Mass Spectrometry

1. Seed Cdc20 knockout cells in 3–5 × 15 cm dishes and synchronize cells by double thymidine arrest as described in above.
2. Wash cells once by 15 mL of PBS and add fresh DMEM medium with RO3306 (5 μM). Incubate cells overnight at 37 °C in CO_2 incubator.
3. Wash cells three times gently by 15 mL of PBS each time and add 15 mL of fresh DMEM medium with nocodazole (200 ng/mL). Incubate cells at 37 °C in CO_2 incubator for 60 min (*see* **Note 6**).
4. Shake off mitotic cells and wash twice with ice-cold PBS. Resuspend cells in ice-cold lysis buffer (1×10^7 cells in 200 μL of lysis buffer containing 10 mM Tris–HCl pH 7.5, 150 mM NaCl, 0.5 mM EDTA, 0.5% NP-40 with protease and phosphatase inhibitors) and incubate on ice for 30 min with extensively pipetting every 10 min. Centrifuge cell lysate at 17,000× *g* for 10 min at 4 °C.
5. Transfer supernatant to a sterile 1.5 mL Eppendorf tube and incubate with Cdc20 antibody (approximately 2 μg of antibody for 500 μg of total protein). The mixture is rotated end-to-end at 4 °C overnight.
6. Add 40 μL of Protein A/G PLUS-Agarose to the mixture. Rotate the mixture end-to-end for another 3 h at 4 °C.
7. Collect the beads by centrifugation at 1000× *g* for 5 min at 4 ° C. Wash beads four times with ice-cold wash buffer (10 mM Tris–HCl pH 7.5, 150 mM NaCl, 0.5 mM EDTA).
8. Resuspend the beads with 40 μL of 2× sample buffer. Biol the beads at 95 °C for 5 min. Centrifuge the mixture at 17,000× *g*

for 10 min. Transfer the supernant to a sterile 1.5 mL Eppendorf tube.

9. Separate immunoprecipitated proteins by SDS-PAGE and send gel slices for mass spectrometry.

3.7 Sensitivity Test of Cdc20 Knockout Cells to Cdc20 RNAi

1. Seed Cdc20 knockout cells at 50% confluency in 6-well plate with DMEM medium containing thymidine (2 mM). Cultivate cells at 37 °C in CO_2 incubator for 24 h.
2. Wash cells once by 2 mL of PBS. Keep cultivating cells at 37 °C in CO_2 incubator for 24 h.
3. Dilute siRNA oligos against luciferase as control or Cdc20 (20 nM) in 250 μL of Opti-MEM (*see* **Note 7**). Add 7.5 μL of Lipofectamine RNAiMax into 250 μL of Opti-MEM. After incubation for 5 min at room temperature, transfer diluted Lipofectamine RNAiMax into diluted siRNA oligo, and mix well by pipetting gently several times. Incubate the mixture at room temperature for 20 min.
4. During the incubation, wash cells with 2 mL of PBS, and add 2 mL of Opti-MEM into each well.
5. Dropwise add siRNA oligo/Lipofectamine RNAiMax complex (500 μL) into the well. Incubate cells for 6 h at 37 °C in CO_2 incubator (*see* **Note 8**).
6. Wash cells with 2 mL of PBS once and add 0.5 mL of trypsin into each well. Swirl the plate several rounds to make sure cells are fully covered by trypsin. Incubate the plate at 37 °C in CO_2 incubator for 3 min. Add 1 mL of fresh DMEM medium into each well. Resuspend cells by pipetting several times. Transfer cell suspension into 15 mL centrifuge tube and centrifuge at 110× *g* for 2 min. Resuspend cell pellet in fresh DMEM medium with thymidine (2 mM). The volume of the DMEM medium used here is adjusted according to the cell confluency (*see* **Note 9**). Transfer around 100,000 of cells into one well of 8-well chamber slide containing 200 μL of fresh DMEM medium with thymidine (2 mM). Put the chamber slide into a 10 cm petri dish and incubate at 37 °C in CO_2 incubator overnight.
7. Wash cells in chamber slide once with PBS and add 0.35 mL of fresh DMEM medium. Incubate cells at 37 °C in CO_2 incubator for 4 h.
8. Setup microscopy for live cell imaging during the incubation. Adjust the imaging parameters to minimize light toxicity (*see* **Note 10**).
9. Replace DMEM medium with 350 μL of L15 medium supplemented with 10% FBS (*see* **Note 11**). Install the chamber slide into stage top incubator.

10. Find the focus plane and start filming. The signals from DIC are recorded every 10 min for 16–20 h.
11. Analyze and plot the filming results with Prism.

3.8 Cdc20 RNAi and Rescue for Mitotic Exit

1. Seed Cdc20 knockout cells in 6-well plate at 50% confluency in DMEM medium with thymidine (2 mM). Cultivate cells at 37 °C in CO_2 incubator for 24 h.
2. Dilute 450 ng of YFP-Cdc20 expression plasmid into 250 μL of Opti-MEM. Add 5 μL of Lipofectamine 2000 into 250 μL of Opti-MEM. After incubation for 5 min at room temperature, transfer diluted Lipofectamine 2000 into diluted siRNA oligo, and mix well by pipetting gently several times. The mixture is incubated at room temperature for 20 min.
3. Wash cells in 6-well plate once by 2 mL of PBS and add 2 mL of Opti-MEM into the well during the incubation.
4. Add the DNA/Lipofectamine 2000 complexes (500 μL) into the well with cells to be transfected in a dropwise manner. Incubate cells at 37 °C in CO_2 incubator for another 6 h.
5. Replace Opti-MEM with fresh DMEM medium. Cultivate cells for another 18 h at 37 °C in CO_2 incubator.
6. Perform RNAi against luciferase or Cdc20 as described in Subheading 3.7.
7. Reseed cells into chamber slide as described in Subheading 3.7.
8. Perform live cell imaging as described in Subheading 3.7 except the signals from DIC, FITC, and or TRITC (in case of mCherry-H3 is co-transfected) channels are recorded every 10 min for 16–20 h.
9. Analyze and plot the filming results Prism (Fig. 3).

4 Notes

1. When designing sgRNA, it is recommended to choose several sgRNAs targeting the first few exons. In our study, three sgRNAs targeting exon 1 and 2 were used to generate Cdc20 knockout cells. Two of them worked well and produced multiple cell lines without detectable Cdc20 examined by Western blot [7].
2. The sgRNA must start with a G for the system we used here. In this case, an extra G is added in the forward primer after CACC and an extra C is added at the 3′ end of the reverse primer.
3. For knockout screening of the clones after puromycin selection, it is important to use antibody against C-terminus of Cdc20. The antigen recognized by the antibody against

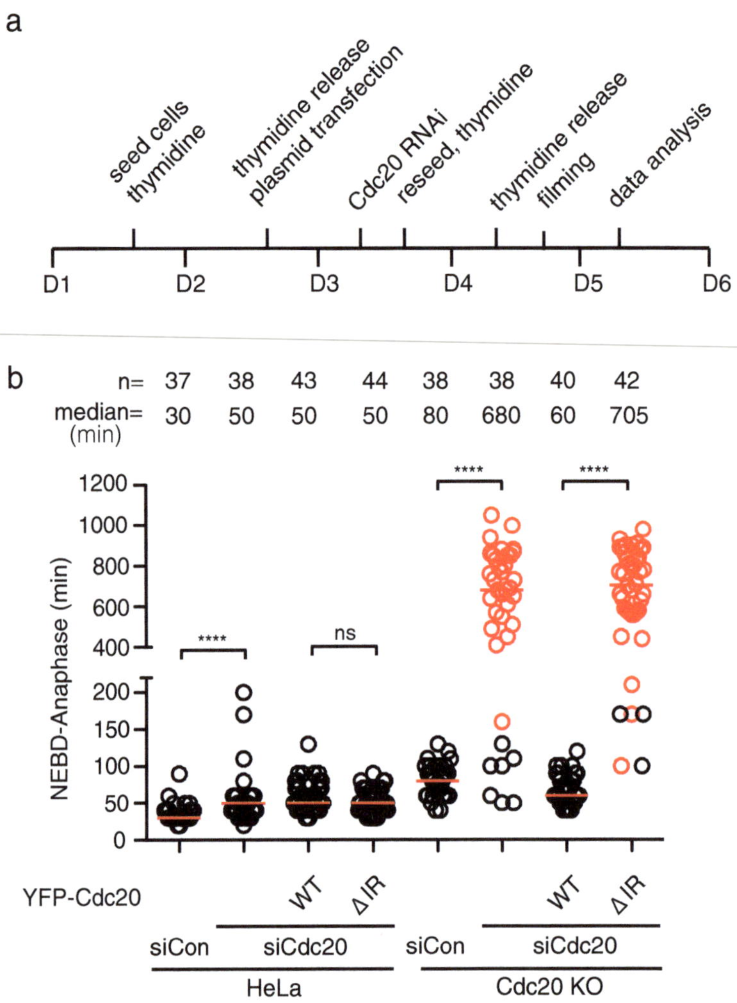

Fig. 3 RNAi and rescue experiment showing the superiority of Cdc20 knockout cells to parental HeLa cells for Cdc20 functionality characterization. (**a**) The strategy of the RNAi and rescue experiment. Cells are transfected with YFP-Cdc20 expression construct 48 h and treated with RNAi against luciferase as control or Cdc20 24 h before functional analysis by live cell imaging. Signals from DIC and YFP are recorded every 10 min for 18 h continuously by the microscopy. (**b**) Results of live cell imaging showing after RNAi treatment, very short arrest occurs to HeLa cells while majority of knockout cells are arrested at metaphase for more than 10 h till apoptosis (*see* **Note 12**). There is no difference of mitotic progression in HeLa cells complemented with exogenous wild type Cdc20 or Cdc20 △IR after depleting endogenous Cdc20. Same treatment in Cdc20 knockout cells display a huge difference between wild type Cdc20 and Cdc20 △IR. Each dot represents the time from nuclear envelope breakdown (NEBD) to anaphase of a single cell. Red dot means cell death after mitotic arrest and the red line indicates the median time. The number of cells analyzed per condition is shown above ($n = x$). Mann-Whitney u-test is applied for statistical analysis. ns means not significant; **** means $P < 0.0001$

N-terminus may be disrupted on the residual Cdc20 protein via alternative splicing or translation reinitiation.

4. The primers used for detecting DNA insertions or deletions (indels) are designed to amplify the genomic DNA between 200 nt upstream and 200 nt downstream of the potential cutting site guided by sgRNA. In case of no PCR product amplified, one may consider to use primers amplifying longer genomic region.
5. Taq DNA polymerase instead of high fidelity DNA polymerase is used here to generate 3′ end overhanging dA for TA cloning.
6. Nocodazole cannot activate SAC in Cdc20 knockout cells. Cells exit mitosis around 90–100 min after mitotic entry even in the presence of nocodazole. For the assay, cells are collected 60 min after releasing from RO3306 arrest. The purpose of using nocodazole here is not to enrich mitotic cells but to achieve a homogenous cell population.
7. RNAi oligo against Cdc20 (5′ CGGAAGACCUGCCGUUA-CAtt 3′) used here targets the middle region of the full-length transcript. In case of lacking sequence information of the transcript for the residual protein, we recommend trying siRNA oligos targeting the middle region or region close to 3′ end of the wild type transcript, as sgRNA normally targets the first few exons and the potential alternative transcript likely has sequence variations close to the 5′ end.
8. RNAi against Cdc20 in knockout cells is performed 24 h before live cell imaging in between the two-thymidine arrest. Since the cells are very sensitive to Cdc20 RNAi, a low concentration of siRNA oligos (20 nM in this study) is enough to deplete almost all the residual protein. Thus, the cells are arrested at metaphase for more than 10 h till apoptosis occurs and very few cells could exit to the next cell cycle.
9. After Cdc20 RNAi, cells are reseeded from 6-well plate into 8-well chamber slide. To make things easier, we normally collect all cells from one well and resuspend cells in 700 μL of DMEM medium if cells reach 100% confluency in the well. 100 μL of cell suspension is seeded into one well of 8-well chamber slide filled with 200 μL of DMEM medium. If the cells are not 100% confluent, scale down the medium volume according to the estimated percentage of confluency. For example, if cells are only 80% confluent, resuspend the cell pellet in 560 μL of DMEM medium and again transfer 100 μL of cell suspension into one well of chamber slide filled with 200 μL of DMEM medium.
10. To minimize toxicity toward cells by repeated exposure to the excitation light during live cell imaging which usually takes around 20 h, we normally use low power input and short

exposure time. In case of confocal microscopy, averaging is set to 1 and low pixel images like 512 × 512 are taken.

11. L15 medium (Leibovitz's L-15 medium) is buffered by phosphates and free base amino acids instead of sodium bicarbonate. Thus, cells cultivated in L15 medium are able to grow without CO_2 equilibration. Do not put chamber slide back to CO_2 incubator once L15 medium is applied. In the live imaging system used here, we only need to set up the incubator temperature at 37 °C and provide water to keep the incubator humid.

12. Optimized Cdc20 RNAi in HeLa cells like using higher concentration of siRNA oligo (50–125 nM) at 36 h prior to live cell imaging could induce slightly longer metaphase arrest (84 min) followed by mitotic exit [8].

Acknowledgments

We thank all the lab members and the National Natural Science Foundation of China (31970666) and Taishan Scholar Project (tsqn201812054) from Shandong, China, for the financial support.

References

1. Kapanidou M, Curtis N, Bolanos-Garcia VM (2017) Cdc20: at the crossroads between chromosome segregation and mitotic exit. Trends Biochem Sci 42:193–205
2. Wolthuis R, Clay-Farrace L, van Zon W et al (2008) Cdc20 and Cks direct the spindle checkpoint-independent destruction of cyclin A. Mol Cell 30:290–302
3. Baumgarten AJ, Felthaus J, Wäsch R (2009) Strong inducible knockdown of APC/C^{Cdc20} does not cause mitotic arrest in human somatic cells. Cell Cycle 8:643–646
4. Li M, York JP, Zhang P (2007) Loss of Cdc20 causes a securin-dependent metaphase arrest in two-cell mouse embryos. Mol Cell Biol 27: 3481–3488
5. Wang Y, Zhai Y, Zhang M et al (2024) Escaping from CRISPR-Cas-mediated knockout: the facts, mechanisms and applications. Cell Mol Biol Lett 29:48
6. Zhang G, Kruse T, Boldú CG et al (2019) Efficient mitotic checkpoint signaling depends on integrated activities of Bub1 and the RZZ complex. EMBO J 38:e100977
7. Zhang Y, Young R, Garvanska DH et al (2024) Functional analysis of Cdc20 reveals a critical role of CRY box in mitotic checkpoint signaling. Commun Biol 7:164
8. Hein JB, Nilsson J (2014) Stable MCC binding to the APC/C is required for a functional spindle assembly checkpoint. EMBO Rep 15:264–272

Chapter 3

Overexpression of GFP Fusions of Regulators of the SAC from *Arabidopsis thaliana*

Charlotte Pain

Abstract

Cell division is a fundamental biological process, essential for sustaining life on Earth. Accurate replication followed by uniform segregation of the genome is required to ensure cell division is sustainable and reduces the likelihood of aneuploidy. The cell cycle has various checkpoints to safeguard proper replication, for example, the spindle assembly checkpoint (SAC) which ensures that all chromosomes are correctly aligned and attached to the spindle, before the transition to anaphase. The precise function of the SAC and SAC components in plants is so far unclear. First, the high level of polyploidy in plants raises concerns about the efficacy of the SAC. Second, many plant SAC components are implicated in other cellular processes, such as MAD1, which has been implicated in the reproductive transition of *Arabidopsis thaliana*. Overexpression of GFP fusions of core SAC components provides a key route to establish the functions of the different SAC components in plants. Here we describe two methods for agrobacterium-mediated transformation of plants.

Key words Spindle assembly checkpoint (SAC), GFP, Agrobacterium, *Arabidopsis thaliana*, Transient expression, Floral dip

1 Introduction

During the process of cell division, the genome of a cell must be accurately replicated and evenly segregated to safeguard genetic identity. During cell division, chromosomes are duplicated, condensed and, after aligning correctly on the mitotic plane, attached to spindle fibers. After this attachment, the chromosomes are separated evenly into two daughter cells. Failure to correctly segregate genetic material between daughter cells results in aneuploidy, which can lead to cancer or cell death in mammalian cells [1]. Early research into cell cycle progression revealed that delaying the arrival of even a single chromosome to the correct position would result in delayed anaphase, which would not commence until several minutes after the arrival of the delayed chromosome [2]. These observations led to speculation that there may be a cell

Victor M. Bolanos-Garcia (ed.), *Mitotic Exit: Methods and Protocols*, Methods in Molecular Biology, vol. 2874,
https://doi.org/10.1007/978-1-0716-4236-8_3,

cycle checkpoint that prevented cell cycle progression until the correct conditions were reached, putatively named the spindle assembly checkpoint (SAC) [3]. Subsequent genetic screening of yeast identified the core SAC components, specifically MAD1, MAD2, MAD3, BUB1, and BUB3 [4, 5] with MPS1 identified later [6].

At the metaphase to anaphase transition, the SAC monitors the attachment of the kinetochore to microtubules. Prior to correct attachment the SAC is "active" and inhibits anaphase by producing the mitotic checkpoint complex (MCC) composed of CDC20, MAD2, BUBR1, and BUB3 [7]. This inhibits the function of the anaphase promoting complex/cyclosome (APC/C), an E3 ubiquitin ligase that controls chromatid separation by opening the cohesion rings that bind sister chromatids. Once the kinetochores are fully attached, the SAC is satisfied and cell division is able to proceed [8, 9].

Thus far, the SAC and its function in cell cycle progression have been most well studied in mammalian cells and yeast, between which the SAC components are well conserved [10]. Much less is known about the function of the SAC in plants, however homologs for the core SAC components and CDC20 have been identified in *Arabidopsis thaliana* (hereafter referred to as *Arabidopsis*), albeit with gene duplication and divergence [11–15]. Though the core components have been identified, the relatively high level of genetic divergence among these genes has raised questions as to their function, and whether they may have alternative functions outside cell cycle control [7]. For example, *Arabidopsis* MAD1 has been implicated in the timing of flowering and therefore the timing of reproductive transition [16]. In addition, polyploidization events are uniquely common in flowering plants, with 30–70% extant plant species characterized as being polyploid [17], leading to the question of whether the SAC acts efficiently in plants [18].

Expression of SAC component GFP fusions in plants is a key step in understanding the structure and function of the plant SAC. Expressing GFP fusions in plants can be achieved in numerous ways, including: biolistic techniques [19], polyethylene glycol (PEG)-mediated transformation of plant protoplasts [20] and microinjection of chloroplasts and nuclei [21]. However, by far the most popular and simple method is *Agrobacterium tumefaciens*-mediated transformation, performed either transiently in plant leaves or stably across the whole plant.

Agrobacterium tumefaciens (hereafter referred to as agrobacterium), is the causative agent of crown gall disease which triggers the formation of tumors on vines, nut trees, perennial fruit trees, and roses [22–24]. Agrobacteria contains a large extrachromosomal element known as the Ti plasmid (tumor-inducing plasmid [25]), which has been used as the basis of numerous recently developed bioengineering processes. In nature, after wounding of the plant, Agrobacteria are able to multiply in the wound and attach to plant cell walls within the wound [26]. Once attached, T-DNA is transferred from the Agrobacteria to the plant cells where it is expressed, even prior to integration [27]. By taking advantage of this system, numerous plant species can be transformed, including tobacco [28], maize [29], potato [30], tomato [31], and rapeseed [32].

Agrobacteria can be used to transform plants either stably or transiently, and which method is most appropriate depends on the experimental requirements. Stable transformation has the advantages of being heritable, with relatively consistent expression levels in different cell types. In addition, transgenic lines can be kept for an extended period of time in the form of transgenic seeds. However, the process of creating a stable transgenic line is laborious and time-consuming. Transient transformations however, allow for the accelerated study of gene localization and function, as it only takes a few days to express target genes once an appropriate construct has been produced. Tobacco is often used as a heterologous transient expression system for *Arabidopsis* protein localization, purification, and co-localization. A large number of studies have used tobacco for multiple experimental purposes and it is typically the case that protein localization and cellular function are similar between *Arabidopsis* and tobacco [33]. A recent study has also shown that, when comparing the localization of 60 cell cycle proteins, the expression patterns were very similar between tobacco and *Arabidopsis* [34]. The same is true of SAC components, for example, MAD2 (Fig. 1).

Here I describe the preparation and methods behind two Agrobacterium-mediated transformation methods for the expression of GFP fusions of regulators of the SAC from *Arabidopsis*:

1. Calcium chloride preparation of Agrobacterium competent cells.
2. Transformation of Agrobacterium.
3. Transient transformation of tobacco leaf epidermal cells.
4. Stable transformation of *Arabidopsis* by floral dip.

2 Materials

2.1 Calcium Chloride Preparation of Agrobacterium Competent Cells

1. *Agrobacterium tumefaciens* (strain: GV3101).
2. Appropriate binary vector.
3. Tryptone 10 g/L, NaCl 10 g/L, Yeast extract 5 g/L in deionized H_2O sterilized with appropriate antibiotics (e.g., rifampicin, kanamycin) to make LB broth (*see* **Note 1**).

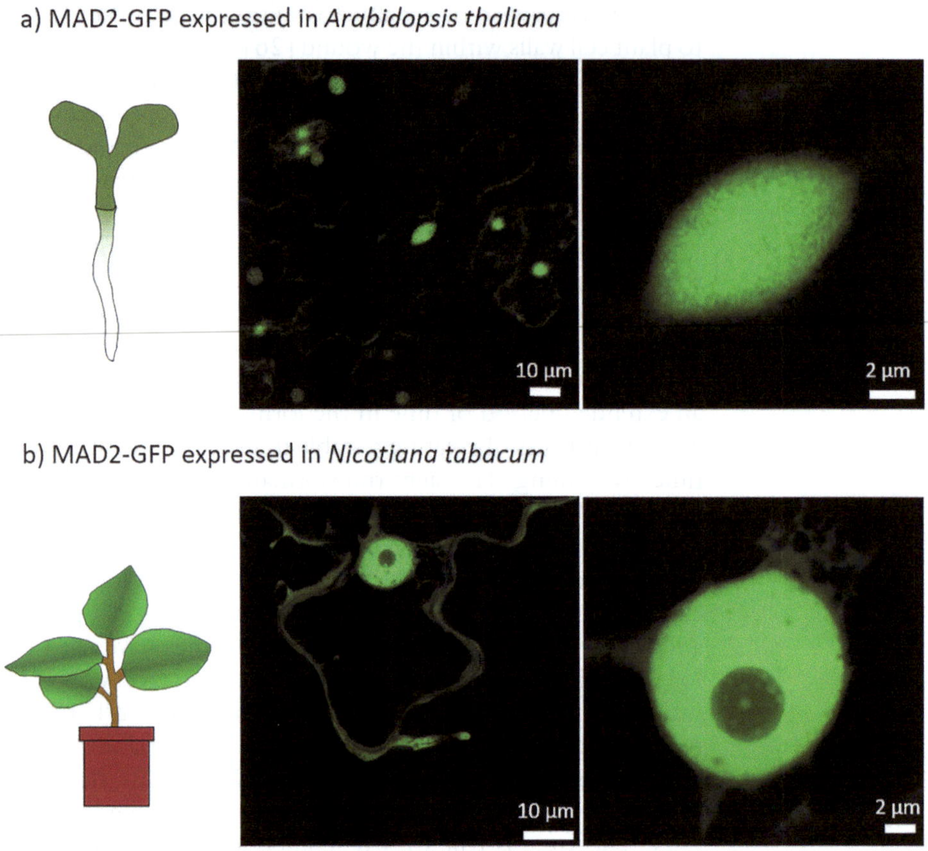

Fig. 1 Comparison of the expression of the plant SAC component MAD2 in *Arabidopsis* and tobacco. (**a**) Stable overexpression of MAD2-GFP in 10-day-old *Arabidopsis* cotyledon epidermal cells alongside and (**b**) transient overexpression of MAD2-GFP in 6-week-old tobacco leaf epidermal cells. In both cases, MAD2-GFP is localized to the nucleus. Images collected using confocal microscopy

4. 20 mM $CaCl_2$, 15% glycerol in deionized H_2O, sterilized to make $CaCl_2$ solution.
5. Conical flasks.
6. Falcon tubes.
7. 1.5 mL Eppendorf safelock tubes
8. Ice box.
9. Centrifuge.
10. Shaking incubator.
11. Nanodrop spectrophotometer (or equivalent).
12. −80°C freezer.

2.2 Transformation of Agrobacterium Cells

1. Competent *Agrobacterium tumefaciens* (strain: GV3101).
2. Peptone 140 10 g/L, NaCl 5 g/L, Yeast extract 5 g/L in deionized H_2O, sterilized with appropriate antibiotics (e.g., rifampicin, kanamycin) to make LB broth.
3. 10 g/L Peptone 140, 5 g/L Yeast Extract, 5 g/L NaCl, 12 g/L Agar in deionized H_2O, sterilized with appropriate antibiotics (e.g., rifampicin, kanamycin) to make LB agar.
4. 1.5 mL or 2 mL Eppendorf tubes.
5. Ice box.
6. Liquid nitrogen in a suitable container, e.g., Dewar.
7. Water bath set to 37 °C.
8. Timer.

2.3 Transient Overexpression of SAC Component GFP Fusions in Tobacco Leaves

1. *Nicotiana tabacum* SR1 cv Petit Havana (hereafter referred to as tobacco) grown in greenhouse conditions to approximately 5 weeks old.
2. 500 mM MES in dH_2O, 20 mM $Na_3PO_4.12H_2O$ (sodium orthophosphate) in dH_2O and 200 mM acetosyringone with DMSO (*see* **Note 2**) to prepare infiltration buffer components.
3. 250 mg D-glucose, 5 mL 500 mM MES, 5 mL 20 mM $Na_3PO_4.12H_2O$, 25 μL 200 mM acetosyringone and make up to 50 mL with deionized H_2O prepared fresh on the day to make infiltration buffer.
4. Peptone 140 10 g/L, NaCl 5 g/L, Yeast extract 5 g/L with appropriate antibiotics (e.g., rifampicin, kanamycin) to make LB broth.
5. 2 mL Eppendorf tubes
6. 1 mL syringes without needle
7. Shaking incubator.
8. Centrifuge.
9. Nanodrop spectrophotometer (or equivalent).
10. Fluorescence microscope.

2.4 Stable Overexpression of SAC Component GFP Fusions in Arabidopsis thaliana

1. *Arabidopsis thaliana* grown in greenhouse conditions to 4–6 weeks old or until flowering.
2. Transformed *agrobacterium tumefaciens.*
3. 5% sucrose and 500ul/L Silwet L-77 in deionized H_2O to make floral dip media.
4. Peptone 140 10 g/L, NaCl 5 g/L, Yeast extract 5 g/L with appropriate antibiotics (e.g., rifampicin, kanamycin) to make LB broth.
5. 2 × 50 mL falcon tubes.

6. Centrifuge.
7. Cling film and growth trays.
8. Shaking incubator.

3 Methods

3.1 Calcium Chloride Preparation of Agrobacterium Competent Cells

Competent Agrobacterium can be prepared quickly and simply using a number of different Agrobacterium strains. Described here is a method to prepare competent Agrobacterium using calcium chloride (*see* **Note 3**):

1. Prepare a culture of *Agrobacterium tumefaciens* (strain GV3101) with appropriate antibiotics in LB media. Grow overnight at 28 °C with shaking.
2. Early the next day, in a conical flask, mix 100 mL LB broth with an appropriate antibiotic for the Agrobacterium strain, and then inoculate with the initial overnight culture.
3. Grow the culture during the day at 28 °C with shaking.
4. Once the culture begins to become visibly thick, monitor the optical density (OD_{600}) of the culture every hour by collecting a small sample and testing using a spectrophotometer.
5. Continue testing the optical density of the culture until it reaches OD_{600} 0.1 (*see* **Note 4**), at which point the culture is ready for the next stage.
6. Transfer the Agrobacterium culture to 2 × 50mL falcon tubes and chill on ice for 15 min.
7. Make a $CaCl_2$ solution composed of sterile 20 mM $CaCl_2$ in 15% glycerol. Chill the solution and ensure it is kept ice cold throughout the process.
8. Pellet the Agrobacterium culture at 3220× *g* for 30 mins at 4 °C.
9. Resuspend each pellet in 15 mL ice cold $CaCl_2$.
10. Pellet the culture again at 3220× *g* for 10 min at 4 °C.
11. Resuspend each pellet in 15 mL ice cold $CaCl_2$ solution.
12. Make aliquots of 400 μL and freeze at −80 °C.

3.2 Transformation of Agrobacterium Cells

Before transforming competent Agrobacterium, create a suitable binary vector for plant transformation. This can be done using multiple techniques and is the subject of several in-depth reviews [35–40] and so will not be covered in detail here. Commonly used vectors for plant transformation include those created by Karimi et al. [37] which uses the Cauliflower mosaic virus (CaMV) 35S promoter and vectors created by Grefen et al. [39], which makes use of the ubiqutin-10 promoter. Both sets of vectors are suitable for overexpression of fluorescent fusion constructs in *Arabidopsis* and tobacco. To transform competent Agrobacterium (stain: GV3101):

1. Take competent Agrobacterium from the −80 °C freezer and allow to completely defrost on ice.
2. Mix approximately 400 ng binary vector with 100 μL competent Agrobacterium (do not vortex) and leave on ice for 5 min in Eppendorf tubes.
3. Rapidly freeze the Agrobacterium mixture in liquid nitrogen.
4. Place the Agrobacterium mixture in water bath set to 37 °C for 4 min (*see* **Note 5**).
5. Add 1 mL LB broth and allow the Agrobacterium to recover in an incubator at 28 °C for 3 h with shaking.
6. Spread the Agrobacterium on selection plates with the appropriate antibiotic.
7. Allow plates to dry and then incubate for 3 days at 28 °C.
8. A single colony can then be picked and grown for transformation (*see* **Note 6**).

3.3 Transient Overexpression of SAC Component GFP Fusions in Tobacco Leaves

Transient expression in tobacco leaves is an efficient way to establish GFP fusion localization.

1. Grow tobacco (*Nicotiana tabacum* SR1 cv Petit Havana) to approximately 5 weeks old in greenhouse conditions.
2. Prepare buffers (*see* **Note** 7), 500 mM MES in deionized H_2O, 20 mM $Na_3PO_4.12H_2O$ in deionized H_2O and 200 mM acetosyringone with DMSO.
3. The day before the infiltration, prepare a 3 mL LB broth overnight Agrobacterium culture with appropriate antibiotics in an incubator at 28 °C with shaking.
4. On the day of the infiltration, prepare the infiltration medium using the buffers made previously (50 mM MES, 2 mM $Na_3PO_4.12H_2O$, 0.1 mM acetosyringone, and 5 mg mL^{-1} glucose).
5. Pellet 2 mL of the overnight culture at 2200× *g* for 5 min at room temperature.
6. Resuspend the pellet in 1 mL infiltration medium.
7. Pellet the agrobacterium suspension at 2200× *g* for 5 min at room temperature.
8. Resuspend the pellet in 1 mL infiltration medium.
9. Measure the bacterial suspension optical density (OD_{600}) using a spectrophotometer and dilute in infiltration medium to an OD_{600} 0.1 (*see* **Note 8**).

10. Using a pipette tip, make a small hole in the underside of a large leaf on the tobacco plant (*see* **Note 9**).
11. Using a 1 mL syringe, collect the bacterial suspension and place the opening of the syringe over the hole. Using a gloved finger to prevent the suspension leaking from the other side of the leaf, use the syringe to push the agrobacterium through the leaf airspaces (Fig. 2).
12. Outline the region in which the Agrobacterium suspension has spread using a large felt tip pen (*see* **Note 10**).
13. Keep the plants in greenhouse conditions for 2–3 days.

Fig. 2 Infiltration of tobacco leaf epidermal leaves. The final stages of Agrobacterium-mediated transient transformation of a tobacco plant by infiltration. (**a**) A healthy *Nicotiana tabacum* grown to 5 weeks in greenhouse conditions. (**b**) Use a pipette tip to make a small hole in the underside of the tobacco leaf. (**c**) Place a 1 mL syringe (with no needle) over the hole and use a finger to block the other side of the leaf. Depress the plunger to push the infiltration buffer through the leaf segment. (**d**) A tobacco leaf with three infiltrated sections. Infiltrated sections can be identified by their slightly lighter coloring. (**e**) Psuedocoloring of three different infiltration "zones." As these areas are separated by leaf veins, each zone can be infiltrated with a different construct without a risk of the constructs mixing, allowing for multiple localization experiments on the same leaf. (**f**) An example of a labeled tobacco leaf, with infiltrated areas drawn around in pen so that they can be identified at a later date

14. Use fluorescence microscopy, for example, laser scanning confocal microscopy, to assess whether the construct has been successfully expressed.

3.4 Stable Overexpression of SAC Component GFP Fusions in Arabidopsis thaliana

An alternative to rapid transient transformation of plant tissue is the creation of stable overexpression line in target plant species, for example, *Arabidopsis.* Detailed here is the floral dip method for stable transformation of *Arabidopsis.*

1. Grow *Arabidopsis thaliana* in greenhouse conditions to 4–6 weeks old (*see* **Note 11**).
2. Prepare a 100 mL culture of transgenic Agrobacterium in a conical flask using LB broth and appropriate antibiotics on the day before the floral dip.
3. Grow the culture overnight at 28 °C with shaking.
4. Transfer the culture to 2 × 50mL falcon tubes and pellet at 3220× *g* at room temperature for 10 min.
5. Prepare the floral dip media.
6. Resuspend the Agrobacterium in the floral dip media, filling each falcon tube up to 20 mL. Combine the two falcon tubes into a single falcon tube containing 40 mL of the Agrobacterium floral dip culture.
7. Gently twist together the stems of the *Arabidopsis* plants to make it easy to dip, and then insert the flowers and stems of the *Arabidopsis* plant into the falcon tube containing the Agrobacterium suspension.
8. Keep the plant submerged for 10–30 s while lightly agitating the mixture to ensure good penetration of the floral dip media into the developing flowers.
9. Remove the plant from the flacon tube and lie it down on a growth tray. Cover the growth tray in cling film for 24 h and keep in greenhouse conditions (*see* **Note 12**). After 24 h remove the cling film and stand the plant back upright (*see* **Note 13**).
10. Repeat the procedure after 5–7 days for enhanced chance of successful transformation.
11. After 2 weeks, stop watering the plants and collect seeds as usual.

4 Notes

1. Other strains of Agrobacteria can be used for this process and some will have different antibiotic requirements. Rifampicin is suitable for the Agrobacterium strain GV3101.

2. MES and sodium orthophosphate buffers can be stored in the fridge, while acetosyringone can be stored in the freezer.
3. It is now possible to buy commercially available competent Agrobacterium with a range of strains and with chemically competent and electro-competent options. However, these commercially available strains can be expensive, whereas "home-made" competent Agrobacterium can be made relatively cheaply.
4. At this stage, the Agrobacterium should be in their log phase of growth.
5. It is important that the sample is left for precisely 4 min, as extending or reducing this time may result in reduced transformation efficiency.
6. Agrobacterium can be stored at −80 °C with 50% sterile glycerol in deionized H_2O for several years.
7. Buffers can be prepared in advance and stored appropriately, *see* **Note 2**.
8. Commonly used optical densities range from OD_{600} 0.05–0.3, for initial testing, an optical density of OD_{600} 0.1 is usually a good starting point.
9. This hole should be made in the center of a leaf segment. Each of the large veins coming off the midrib of the leaf can be used to make separate "infiltration zones." Each of these zones can be infiltrated with a different construct as the infiltration buffer will not spread beyond the vein, allowing for multiple localization experiments on the same leaf.
10. This step is essential as by the time the plant is ready for imaging, it will be almost impossible to distinguish areas that have been infiltrated with Agrobacterium from the surrounding leaf tissue. Avoid damaging the leaves as much as possible while marking the boundaries of the infiltration area. In addition, be aware that some marker pens may fluoresce on a fluorescence microscope.
11. For best transformation efficiencies, it is best to cut back the primary bolts and then use the secondary bolts for transformation. For highest transformation efficiency, it is recommended that the secondary bolts are 2–10 cm in length with few open flowers. After the development of numerous mature siliques, the transformation efficiency drops rapidly [41].
12. The goal is to increase humidity around the plant to enhance agrobacterium survival.
13. It is essential to remove the cling film in a timely manner as the plant health can be reduced in high humidity conditions.

Acknowledgments

This work was supported by an Early Career Fellowship grant from the Leverhulme Trust (ECF-2022-002).

References

1. Epstein CJ (1988) Mechanisms of the effects of aneuploidy in mammals. Annu Rev Genet 22: 51–75
2. Zirkle RE (1970) Ultraviolet-microbeam irradiation of newt-cell cytoplasm: spindle destruction, false anaphase, and delay of true anaphase. Radiat Res 41(3):516–537
3. Hartwell LH, Weinert TA (1989) Checkpoints: controls that ensure the order of cell cycle events. Science 246:629–634
4. Hoyt MA, Totis L, Roberts BT (1991) S. cerevisiae genes required for cell cycle arrest in response to loss of microtubule function. Cell 66:507–517
5. Li R, Murray AW (1991) Feedback control of mitosis in budding yeast. Cell 66:519–531
6. Weiss E, Winey M (1996) The Saccharomyces cerevisiae spindle pole body duplication gene MPS1 is part of a mitotic checkpoint. J Cell Biol 132:111–123
7. Komaki S, Schnittger A (2016) The spindle checkpoint in plants – a green variation over a conserved theme. Curr Opin Plant Biol 34: 84–91
8. Etemad B, Kuijt TEF, Kops GJPL (2015) Kinetochore–microtubule attachment is sufficient to satisfy the human spindle assembly checkpoint. Nat Commun 6:8987
9. Tauchman EC, Boehm FJ, DeLuca JG (2015) Stable kinetochore–microtubule attachment is sufficient to silence the spindle assembly checkpoint in human cells. Nat Commun 6: 10036
10. London N, Biggins S (2014) Signalling dynamics in the spindle checkpoint response. Nat Rev Mol Cell Biol 15:736–748
11. Lermontova I, Fuchs J, Schubert I (2008) The Arabidopsis checkpoint protein Bub3.1 is essential for gametophyte development. Front Biosci 13:5202–5211
12. Caillaud MC, Paganelli L, Lecomte P et al (2009) Spindle assembly checkpoint protein dynamics reveal conserved and unsuspected roles in plant cell division. PLoS One 4:e6757
13. Marrocco K, Thomann A, Parmentier Y et al (2009) The APC/C E3 ligase remains active in most post-mitotic Arabidopsis cells and is required for proper vasculature development and organization. Development 136(9): 1475–1485
14. Kevei Z, Baloban M, Da Ines O et al (2011) Conserved CDC20 cell cycle functions are carried out by two of the five isoforms in Arabidopsis thaliana. PLoS One 6(6):e20618
15. de Oliveira EA, Romeiro NC, Ribeiro Eda S et al (2012) Structural and functional characterization of the protein kinase Mps1 in Arabidopsis thaliana. PLoS One 7:e45707
16. Bao Z, Zhang N, Hua J (2014) Endopolyploidization and flowering time are antagonistically regulated by checkpoint component MAD1 and immunity modulator MOS1. Nat Commun 5:5628
17. Ramsey J, Schemske DW (1998) Pathways, mechanisms, and rates of polyploid formation in flowering plants. Ann Rev Ecol Evol Syst 29: 467–501
18. Hollister JD (2015) Polyploidy: adaptation to the genomic environment. New Phytol 205: 1034–1039
19. Lacroix B, Citovsky V (2020) Biolistic approach for transient gene expression studies in plants. Methods Mol Biol 2124:125–139
20. Junker B, Zimny J, Luhrs R et al (1987) Transient expression of chimaeric genes in dividing and non-dividing cereal protoplasts after PEG-induced DNA uptake. Plant Cell Rep 6: 329–332
21. Knoblauch M, Hibberd JM, Gray JC (1999) A galinstan expansion femtosyringe for microinjection of eukaryotic organelles and prokaryotes. Nat Biotech 17:906–909
22. Escobar MA, Leslie CA, McGranahan GH (2002) Silencing crown gall disease in walnut (*Juglans regia L.*). Plant Sci 163:591–597
23. Aysan Y, Sahin F (2003) An outbreak of crown gall disease on rose caused by *Agrobacterium tumefaciens* in Turkey. Plant Pathol 52:780–780
24. Chen F, Guo YB, Wang J et al (2007) Biological control of grape crown gall by *Rahnella aquatilis* HX2. Plant Dis 91:957–963

25. van Larebeke N, Genetello C, Schell J et al (1975) Acquisition of tumour-inducing ability by non-oncogenic agrobacteria as a result of plasmid transfer. Nature 255:742–743
26. Lippincott BB, Lippincott JA (1969) Bacterial attachment to a specific wound site as an essential stage in tumor initiation by agrobacterium tumefaciens. J Bateriol 97:620–628
27. Janssen BJ, Gardner RC (1989) Localized transient expression of GUS in leaf discs following cocultivation with agrobacterium. Plant Mol Biol 14:61–72
28. Horsch RB, Fry JE, Hoffmann NL et al (1985) A simple and general method for transferring genes into plants. Science 227:1229–1231
29. Ishida Y, Saito H, Ohta S et al (1996) High efficiency transformation of maize (*Zea mays* L.) mediated by *Agrobacterium tumefaciens.* Nat Biotechnol 14:745–750
30. Stief A, Winter DM, Stratling WH et al (1989) A nuclear DNA attachment dement mediates elevated and position-independent gene activity. Nature 341:343–345
31. Van Eck J, Keen P, Tjahjadi M (2019) Agrobacterium tumefaciens-mediated transformation of tomato. Methods Mol Biol 1864:225–234
32. Charest PJ, Holbrook LA, Gabard J et al (1988) Agrobaeterium mediated transformation of thin cell layer explants from Brassica napus L. Theor Appl Genet 75:438–445
33. Zhang J, He S (2021) Tobacco system for studying protein colocalization and interactions. In: Yin R, Li L, Zuo K (eds) Plant photomorphogenesis. Humana, New York
34. Boruc J, Mylle E, Duda M et al (2010) Systematic localization of the arabidopsis core cell cycle proteins reveals novel cell division complexes. Plant Physiol 152(2):553–565
35. von Arnim AG, Deng X-W, Stacey MG (1998) Cloning vectors for the expression of green fluorescent protein fusion proteins in transgenic plants. Gene 221(1):35–43
36. Goodin MM, Dietzgen RG, Schichnes D et al (2002) pGD vectors: versatile tools for the expression of green and red fluorescent protein fusions in agroinfiltrated plant leaves. Plant J 31(3):375–383
37. Karimi M, Inzé D, Depicker A (2002) GATEWAY™ vectors for agrobacterium-mediated plant transformation. Trends Plant Sci 7(5): 193–195
38. Zhong S, Lin Z, Fray RG et al (2008) Improved plant transformation vectors for fluorescent protein tagging. Transgenic Res 17(5):985–989
39. Grefen C, Donald N, Hashimoto K et al (2010) A ubiquitin-10 promoter-based vector set for fluorescent protein tagging facilitates temporal stability and native protein distribution in transient and stable expression studies. Plant J 64(2):355–365
40. Han J, Ma K, Li H et al (2022) All-in-one: a robust fluorescent fusion protein vector toolbox for protein localization and BiFC analyses in plants. Plant Biotechnol J 20(6):1098–1109
41. Clough SJ, Bent AF (1998) Floral dip: a simplified method for agrobacterium-mediated transformation of Arabidopsis thaliana. Plant J 16(6):735–743

Chapter 4

Production of SAC Proteins in a Baculovirus-Based Heterologous Gene Expression System

Victor M. Bolanos-Garcia

Abstract

The overproduction of certain protein components of the Spindle Assembly Checkpoint (SAC) such as human BUB3, CDH1, and CDC20 either in isolation or associated with other proteins in prokaryotic heterologue gene expression systems remains a technical challenge. Recent advances in virus engineering methods and the development of genetically modified insect cells have been exploited to overproduce these SAC proteins as natively folded, functional molecules. One widely used baculovirus-based expression system is flashBac, which exhibits several advantages over other insect cells-based expression systems such as the straightforward production of recombinant baculoviruses through homologous recombination and the positive selection of recombinant clones.

Key words Spindle Assembly Checkpoint (SAC), Insect cells heterologous gene expression, Protein overproduction, Baculovirus, BUB3, CDH1, CDC20, Flashback system

1 Introduction

Insect cells are a widely used system for the overproduction of recombinant proteins from different species including *Homo sapiens*. This heterologous gene expression system is suitable to overproduce proteins that harbor posttranslational modifications such as phosphorylation and glycosylation [1, 2]. Previous reports have shown that baculovirus-based expression systems enable the production of natively folded SAC proteins [3–5] including most protein subunits of the ultimate SAC effector, the Anaphase Promoting Complex, also commonly known as the Cyclosome (APC/C). This achievement has been instrumental to understand the structural basis of SAC signaling and the control of chromosome segregation with an unprecedented level of detail [6].

The viruses that primarily infect insect larvae of the order Lepidoptera, a taxonomic order that includes moths and butterflies, are known as baculovirus [7] and these are widely use for the

Victor M. Bolanos-Garcia (ed.), *Mitotic Exit: Methods and Protocols*, Methods in Molecular Biology, vol. 2874, https://doi.org/10.1007/978-1-0716-4236-8_4,

overproduction of recombinant proteins in certain types of insect cells. Typically, this procedure starts with the genetic modification of the baculovirus via the insertion of the foreign gene(s) of interest under the control of a baculovirus gene promoter [8]. The heterologous gene is then expressed in a host insect cell line. The budded virus (BV) of the multicapsid nucleopolyhedrovirus (AcMNPV) is the most extensively used system for heterologous gene expression in tissue culture. AcMNPV is a double-stranded, supercoiled DNA genome of approximately 130 kb in length. The genome is packaged into a rod-shaped nucleocapsid [9, 10]. Because the large size of the AcMNPV genome makes it unsuitable for direct foreign gene insertion, the heterologous gene of interest is typically cloned into a transfer plasmid with sequences flanking the *polh* gene in the virus genome. The co-transfection of the virus genome and the transfer plasmid into the host insect cells enable the insertion of the gene of interest into the viral genome at the *polh* locus via homologous recombination. The virus genome is replicated and the recombinant baculovirus gradually generated. Insect cell lines derived from *Spodoptera frugiperda* (*Sf9* and *Sf21*) and *Trichoplusia ni (T.ni)* are the most widely used for gene expression in a baculovirus-based system. These lepidopteran insect cell lines can grow in suspension or adherent cell culture medium (either supplemented with serum or serum-free) containing phosphates buffer to adjust the pH. Sf*9*, Sf*21*, and *T.ni* all grow optimally at 28 °C without the requirement of CO_2, turning the scaling-up of protein overproduction easy and cost-effective [11].

The flashBAC™ baculovirus expression system enables the rapid production of recombinant baculoviruses. The flashBAC™ technology is based on the insertion of a *lacZ* gene into the *polh* gene locus of AcMNPV, in which a Bsu361 restriction enzyme recognition sequence (e.g., 5′-CCTNAGG-3′) flanks the *lacZ* gene [12]. A second hallmark feature of flashBAC™ is the incorporation of a Bacterial Artificial Chromosome (BAC) at the *polh* gene locus and a deletion within the gene *ORF1629*, which is essential for viral replication. Homologous recombination between ORF603 and ORF1629 in flashBAC™ DNA and the transfer plasmid results in the replacement of the BAC replicon with the gene of interest under the polh promoter, while simultaneously restoring ORF1629 gene function. Ultimately, these molecular events result in the replication of the recombinant virus harboring the heterologous gene of interest (Fig. 1). This manner, the production of recombinant virus is achieved in a single step, turning this technology amenable to high-throughput gene expression platforms [13].

One compatible transfer plasmid for flashBAC is the pOET5 vector (Oxford Expression Technologies, OET) (Fig. 2). We used

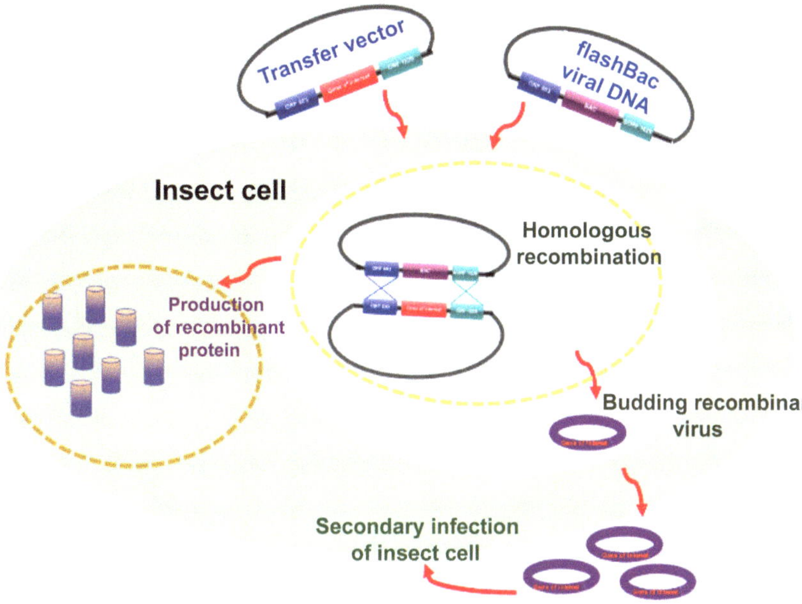

Fig. 1 Schematic diagram of the flashBAC™ technology. The modified AcMNPV genome contains a BAC at *polh* gene locus with a segment of the *ORF1629* gene deleted. The technology involves the co-transfection of insect cells with *flash*BAC DNA and the transfer vector containing the heterologous gene of interest. Restoration of gene function and insertion of the foreign gene into the virus DNA under *polh* promoter occurs by homologous recombination. Baculovirus can be produced as long as the recombinant virus continues to replicate. Baculovirus can be easily harvested by centrifugation and used to infect a new batch of insect cells for the production of the recombinant protein of interest

recently this dual promoter baculovirus vector to overproduce one central protein component of the SAC from *A. thaliana* in *Sf9* and *T. ni* cells [14]. pOET5 allows plasmid propagation and selection in *E. coli* cells due to the incorporation of a bacterial origin of replication and an ampicillin resistance gene. This dual promoter enables the concomitant expression of two foreign genes under the strong AcMNPV *polh* promoter and the very late promoter *p10*. To this aim, two multiple cloning sites (MCS) replace the *polh* and *p10* sequences with unique restriction sites, enabling the insertion of up to two heterologous genes of interest in the correct orientation for their expression in insect cells.

2 Materials

All solutions are prepared at room temperature using ultrapure, deionized water (e.g., purified water with an electrical conductivity of 18 MΩ-cm at 25 °C) unless indicated otherwise. All the reagents required are of analytical grade quality.

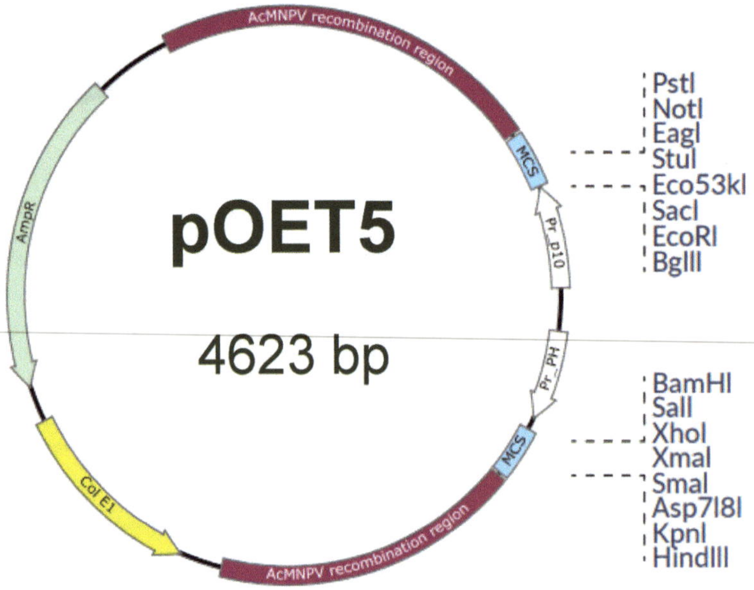

Fig. 2 Schematic representation of the pOET5 dual vector. This transfer vector has a Col E1 bacterial origin of replication and an ampicillin resistant gene that enables selection in the eubacteria *E. coli*. Two multiple cloning sites (MCS) in which a segment of the polyhedrin (*polh*) and *p10* sequences have replaced to insert unique restriction sites, enable the molecular cloning of two heterologous genes in the correct 5′-3′ orientation and open reading frame (ORF). The *Autographa californica* Multicapsid Nucleopolyhedrovirus (AcMNPV) gene sequences flanking the heterologous genes in the transfer vector MCS enable homologous recombination with the viral DNA in insect cells, which in turn facilitates the insertion of a heterologous gene expression cassette into the locus

2.1 SDS (Sodium Dodecyl Sulphate) Polyacrylamide Gel Electrophoresis (SDS-PAGE)

1. 4–12% NuPAGE pre-cast gels 1.5 mm, 15 wells.
2. One thank suitable for NuPAGE.
3. SDS-PAGE running buffer: 50 mM MOPS, 50 mM Tris base, 3.46 mM SDS, 1 mM EDTA (*see* **Note 1**).
4. A protein molecular weight ladder covering the 200–10 KDa range.
5. Leammli buffer 5 × 125 mM Tris–HCl pH 6.8, 10% (v/v) glycerol, 4% (w/v) SDS, 2% (v/v β-mercaptoethanol), 0.02% (w/v) bromophenol blue.

2.2 Immunoblotting

1. Nitrocellulose membranes.
2. Western blot transfer buffer: 25 mM Tris–HCl, 192 mM glycine, 0.01% SDS, 20% (v/v) methanol.
3. PBS buffer: 150 mM NaCl, 50 mM Na phosphates, pH 7.4.
4. PBST buffer: PBS containing 0.05% (v/v) Tween-20.
5. Blocking solution: 5% (w/v) milk in PBST. Store at 4 °C.

6. Thick blot filter paper for Western blot, 9.5 × 15.2 cm.
7. Alkaline Phosphatase (AP) or horseradish peroxidase (HRP) anti-6× His tag antibodies. Secondary antibody that is HRP- or AP-conjugated.
8. Nitro blue tetrazolium (NBT)/5-bromo-4-chloro-3- indolyl phosphate (BCIP) 1-Step™ ready-made mix or Clarity Western ECL ready-made mix.

2.3 Insect Cells Work

1. flashBAC™ kit.
2. pOET5 vector.
3. baculoFECTIN II transfection reagent.
4. SF 921 protein-free cell culture medium.
5. Serum free TC100 medium.
6. 35 mm (six well) plates for tissue culture.
7. 5-bromo-4-chloro-3-indolyl-β-D galactopyranoside (X-gal).
8. N,N Dimethylformamide (DMF).
9. Fetal Bovine Serum (FBS).
10. Penicillin/streptomycin mix (10,000 IU and10000μg/mL respectively).
11. PBS buffer: 150 mM NaCl, 50 mM Na phosphates, pH 7.4.
12. Neutral Red Dye solution diluted (1:20) in sterile Phosphate-buffered saline (PBS).

3 Methods

3.1 Cloning of Heterologous SAC Genes

1. Add a 5′ sequence encoding for a hexahistidine-tandem after the start codon of the SAC heterologue gene sequence to facilitate the purification of the gene expression product by immobilized metal affinity chromatography (IMAC). References [15, 16] present a revision of the chemical principles and applications underpinning IMAC.
2. After gene amplification by PCR, the amplicon of interest is purified and cloned into the appropriate restrictions sites of the pOET5 vector using conventional molecular cloning protocols.
3. Confirmation of appropriate cloning of the heterologous gene (e.g., in-frame, mutations-free) in pOET5 by DNA Sanger sequencing using appropriate pOET sequencing primers.

3.2 Insect Cells Transfection

A seed stocks (P0) of recombinant baculovirus is generated after co-transfection of insect cells. The protocol shown below is carried out under aseptic conditions using standard tissue culture techniques.

1. Grow *Spodoptera frugiperda 9* (*Sf9*) cells in TC100 medium (serum-free) using shake flasks (28 °C, 120 rpm).
2. At least 1 hour prior to transfection, seed 35 mm cell culture dishes with 2 mL of Sf9 cells in serum free TC100 growth medium to reach a cell density of 0.5×10^6 cells/mL. It is very important to use *Sf9* cells that are in the log phase of growth.
3. Incubate the Sf9 cells at 28 °C for 1 h.
4. While the Sf9 cell are incubated at 28 °C, the co-transfection mix of DNA and transfection reagent is prepared. Use sterile tubes for each co-transfection. Add first 100 μL of TC100 medium (serum-free), followed by 100 ng virus DNA supplied in the *flash*BAC™ kit (i.e., use 5 μL of DNA at a concentration of 20 ng/μL).
5. Add 500 ng of transfer vector containing the heterologous gene of interest. In the case of lacZ positive control plasmid, add 500 ng of the material supplied with the *flash*BAC™ kit (i.e., add 5 μL of control plasmid at a concentration of 100 ng/μL).
6. Add 1.2 μL of baculoFECTIN and gently mix to prevent shearing of the DNA.
7. Prepare a control (mock) transfection by omitting the addition of DNA to the mix.
8. Allow the nanoparticle-DNA complex to form by incubating the transfection mix for 15 minutes at room temperature.
9. Gently remove 1 mL of growth medium from each well of the 6-well plates and discard. This will result in each well containing 1 mL of serum free TC100 medium.
10. Add dropwise the co-transfection mix into the corresponding well, ensuring an even distribution of the mixture over the well.
11. Incubate the plates overnight at 28 °C.
12. Following overnight incubation, add 1 mL of ESF 921 protein-free cell culture medium so each well contains a total medium volume of 2 mL.
13. Incubate the 6-well plates at 28 °C for 4 days.
14. Recover the seed stock (P0) fraction from each well and transfer into sterile tubes. Store the P0 fractions in the dark at 4 °C.
15. For the lacZ positive control used to confirm the efficiency of transfection, stain the infected insect cells with X-gal. For this, add 1 mL of ESF 921 protein-free cell culture medium containing 15 μL X-gal (2% w/v prepared in DMF) to the positive control well and incubate at 28 °C for at least 5 h.

16. The gradual formation of a blue-light green color is confirmation of lacZ gene expression in the recombinant virus.

3.3 Baculovirus Amplification

1. The procedure described below is carried out under aseptic conditions.
2. A Passage 1 (P1) virus stock is generated by taking 0.5 mL of the P0 stock virus to infect a volume of 50 mL of Sf9 cells at a cell density of 2×10^6 cells/mL. Use a conical flask of 250 mL capacity to ensure the good aeration of the cells.
3. Incubate the cells for 5 days at 28 °C in a shaker incubator set at 120 rpm.
4. Harvest the insect cells by centrifugation (20 min, 4 °C, 3000 rpm). Recover the supernatant and store in the dark at 4 °C. For long-term storage of the virus P1 stock at −80 °C, add 5% (v/v) of Fetal Bovine Serum (FBS) prior to freezing the cells.
5. Determine the titer of the virus P1 stock by the plaque assay method.

3.4 Plaque Assay

1. The procedure described below is carried out under aseptic conditions.
2. 12-well or 6-well plates are seeded with either Sf9 in ESF 921 protein-free cell culture medium or *Spodoptera frugiperda* 21 (Sf21) cells in TC100 medium supplemented with 10% (v/v) FBS. The recommended cell density for each insect cell line and plate type is shown below (Table 1):
3. Plates are incubated overnight at 28 °C to reach the desired confluency (i.e., ≥70%).
4. Serial dilutions (1:10) of virus stock are performed by tenfold steps (10^{-1}–10^{-7}) in TC100 medium supplemented with 10% (v/v) FBS. For this, use 50 μL of virus and 450 μL of growth medium as diluent at each step, ensuring to mix thoroughly each time before moving to the next serial dilution.

Table 1
Recommended cell density for the plaque assay using different insect cell lines and format

Insect cell line	Plate type	Cell density per well (cells/mL)	Volume per well (mL)
Sf9	6-well	0.35×10^6	2
Sf9	12-well	0.4×10^6	1
Sf21	6-well	0.75×10^6	2
Sf21	12-well	0.4×10^6	1

5. The medium is aspired from the wells using sterile pipettes and 100 μL of virus dilutions gently added dropwise to the center of each well.
6. Plates are incubated for 45 min at room temperature with mild agitation using a rocking shaker to allow the virus to adsorb. In the meantime, an overlay is prepared as follows:
7. Low melting temperature agarose at a concentration of 2% (w/v) prepared in in deionized water is warmed to 50–55 °C and diluted 1:2 with pre-warmed (28 °C) TC100 medium supplemented with 10% (v/v) FBS or ESF 921 protein-free cell culture medium if using *Sf21* or *Sf9* cells, respectively. Keep the mix in an oven set at 55 °C to prevent the agarose gel to set.
8. After the incubation period is over, the virus suspension is aspired and 1 mL or 2 mL of overlay slowly added to 12-well or 6-well plates respectively, allowing the overlay to flow down the side of the well and to spread over the cell monolayer.
9. Once the agarose has set, 1 mL of antibiotic mix (10,000 IU penicillin/10000 μg/mL streptomycin) in TC100 medium +10% FBS or ESF 921 protein-free cell culture medium (if using *Sf21* or *Sf9* cells, respectively) is added to the cells and the plate(s) incubated at 28 °C for 3/4 days for *Sf21/Sf9* cells, respectively.
10. For visual inspection of the virus plaques, the cells are stained with Neutral Red Dye solution diluted (1:20) in sterile PBS. This staining solution must be sterilized prior to use.
11. Plates are incubated overnight at 28 °C, then dye is decanted and plaques counted. The virus titer is determined from the following equation:

$$\text{Average plaque count} \times \text{Dilution factor} \times 10 = \text{Virus titer (pfu/mL)}$$

Where dilution factor is the inverse of the dilution and the multiplication by 10 is because 0.1 mL of virus was added to each well.

3.5 Time-Course Protein Production

Once a virus titer of high infectivity (typically 10^7 to 10^8 pfu/mL) is obtained, time-course expression tests are conducted in insect cells such as Sf9 at different multiplicity of infection (MOI) titers for up to 96 h post infection (Fig. 3).

1. For small-scale production trials of heterologue SAC proteins, the cell lines Sf9 and T.ni at a cell density of 1.5×10^6 cells/mL and 1.0×10^6 cells/mL, respectively, are infected with the P1 virus stock at 0.1, 1.0 and 5.0 Multiplicity of Infection (MOI) titers. Samples of 1 mL from the different MOIs are collected after 24, 48, 72, and 96 h and harvested by centrifugation at

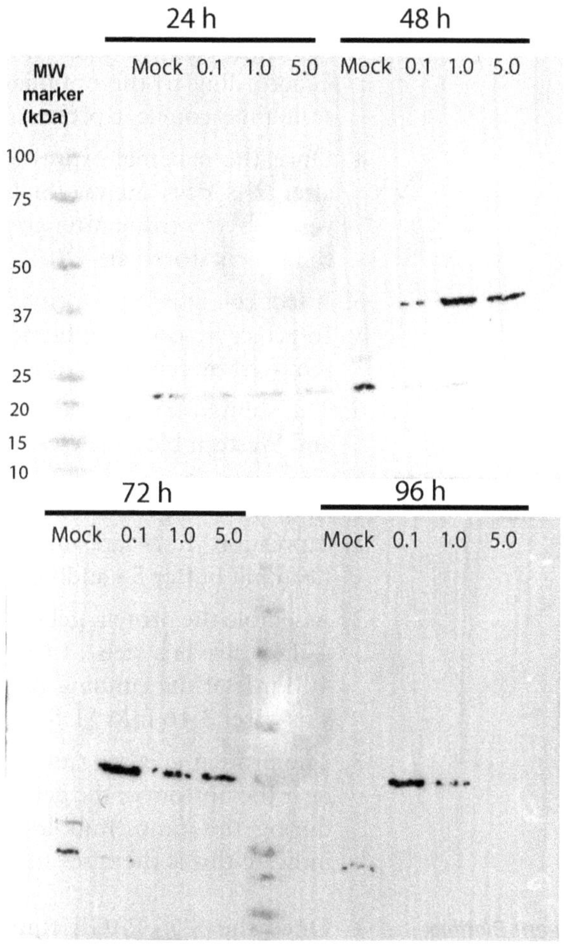

Fig. 3 Expression of a SAC protein in insect cells. The image shows Western blots to monitor the time-course expression of Cdc20, a central component of the Spindle Assembly Checkpoint (SAC) in Sf9 cells. Three different multiplicity of infection (MOI) were assessed (0.1, 1.0 and 5.0). Non-infected Sf9 cells (e.g., mock) were used as a negative control of protein overexpression

12,000 rpm for 10 min. The pellet and supernatant can be stored at −20 °C for further analyses. For insect cells lysis, a lysis buffer solution consisting of 50 mM Tris–HCl pH 8.0, 5 mM β-mercaptoethanol, 100 mM KCl, 1% (v/v) Nonidet P-40, cOmplete™, EDTA-free protease inhibitor cocktail tablets (Roche) is used (*see* **Note 2**).

2. Centrifuge at 12,000 rpm, at 2–8 °C for 15 min to remove cells membrane fragments and other debris. An aliquot of the supernatant (e.g., the soluble fraction) and the pellet (the insoluble fraction) are saved for SDS-PAGE and Western Blotting analysis.

3. For large-scale expression, 1 to 4 L of Sf9 or T.ni insect cells in culture at 1.5×10^6 cells/mL are infected and incubated at 28 °C according to the optimized conditions found in the small-scale time-course expression trials.
4. Once the maximal expression hours are determined (usually, after 2–3 days incubation post-infection), the cells are harvested by centrifugation at 4000 rpm, 2–8 °C for 20 min and the pellets stored at −20 °C.
5. Insect cell lysis is performed as described above (see point 1). Intact cells, cell membrane fragments, and other debris are removed by centrifugation at 12,000 rpm, 2–8 °C for 45 min.
6. The soluble fractions are recovered and used for SDS-PAGE and Western blot analyses.

3.6 SDS-PAGE

1. Aliquots of the collected soluble fractions from the time-course expression tests are mixed with SDS-PAGE loading dye Leammli buffer 5× and heated at 95 °C for 10 min.
2. Assemble the protein gel system with 15-well precast NuPAGE 4–12% Bis-Tris gels (1.0 mm thick). Fill the gel tank with 400 mL of the running buffer (e.g., 50 mM MOPS, 50 mM Tris base, 3.46 mM SDS, 1 mM EDTA).
3. The protein gels are run at 120 V until the dye front reaches near the bottom of the gel (~50 min). Ensure the gel is running during this time. Bubbles coming off the bottom electrode indicate this is the case (*see* **Note 3**).

3.7 Western Blotting

1. Once the SDS-PAGE run is completed, rinse the gel with deionized water and transfer to a small plastic box containing enough transfer buffer solution to cover the gel.
2. Cut a piece of polyvinylidene fluoride (PVDF) membrane of a size similar to that of the protein gel (*see* **Note 4**).
3. Soak the PVDF membrane in pure methanol for 1 min (*see* **Note 5**).
4. Transfer the membrane to a small plastic box containing enough transfer buffer solution to cover the PVDF membrane and incubate for 5 min (*see* **Note 6**).
5. During the equilibration of the PVDF membrane in the transfer buffer, assemble the blotting sandwich as follows: Whatman filter paper-membrane-gel-Whatman filter paper. Place the gel on the PVDF membrane in such a way that its cover the membrane completely. Use a 5 mL plastic pipette to roll out air bubbles from the gel-membrane sandwich prior to placing in transfer cassette. This will prevent the trapping of air bubbles in between the PVDF membrane and the gel.

6. Use a semi-dry transfer blot to transfer the proteins from the gel to the PVDF membrane. Typically, this step performed at 25 V for 5 min per protein gel.
7. Following transfer, the membrane is blocked overnight at 4 °C with rotation in blocking solution consisting of 5% (w/v) skim milk powder in PBST buffer.
8. The primary antibody is diluted according to supplier's recommendations in blocking solution (typically between 1:200 and 1:1000). The membrane is incubated with the primary antibody diluted in PBST for 1–2 h at 4 °C with rotation. The use of Alkaline Phosphatase (AP) or horse radish peroxidase (HRP) anti-6× His tag antibody is recommended to speed up the process.
9. The primary antibody is removed by six 5-min washes with PBST followed by a 1–2 h incubation (4 °C with rotation) with secondary antibody that is HRP- or AP-conjugated (the dilution is according to supplier's recommendations) in PBST.
10. Following six 5-min washes as in **step 9** above, for HRP-conjugated antibody, the colorimetric 1-Step NBT/NCIP ready-made mix is applied to the membrane for the detection of proteins on the blots. For alkaline phosphatase-conjugated antibody use the 1-Step™ NBT/BCIP ready-made mix (*see* **Note 7**).
11. The blot images are captured on a Western blot and gel documentation instrument.

4 Notes

1. Unpolymerized polyacrylamide is a neurotoxin, so you must wear gloves during the handling of SDS-PAGE gels.
2. Insect cell lysis solution can irritate the skin, mucous membranes and eyes. In case of accidental contact, rinse cautiously with water for several minutes.
3. The electrophoresis and transfer apparatus require the use of high voltage.
4. Avoid touching the PVDF membrane with bare fingers as the proteins left on the membrane (mainly keratin), may cross-react with the antibodies.
5. Methanol is a flammable and harmful solvent that may cause blindness or death if swallowed. Store in a well-ventilated, cool area. Use PPE-based containers to handle it. Keep away from strong oxidizing agents. Avoid heat/ignition sources.
6. During the Western blotting procedure, ensure the PVDF membrane, the protein gel and the filter paper sheets are kept

damp. Pouring a small volume of transfer buffer onto the stack to keep it moist during assembling it should be sufficient. It is also essential that there are no air bubbles between any of the stack layers, otherwise protein gel to membrane transfer will not occur.

7. The NBT/BCIP staining solution is a flammable liquid and vapor. Harmful in contact with skin. Causes serious eye irritation.
8. The Tris-Glycine-Methanol blotting buffer solution must be disposed of as hazardous waste.
9. Western blotting bands fade away when stored for a few days even when the membrane is covered with water. Therefore, it is best to image the blots immediately after performing the chemiluminescence reaction.

References

1. Hong M, Li T, Xue W, Zhang S, Cui L, Wang H, Zhang Y, Zhou L, Gu Y, Xia N, Li S (2022) Genetic engineering of baculovirus-insect cell system to improve protein production. Front Bioeng Biotechnol 10:994743
2. Cox MMJ (2021) Innovations in the insect cell expression system for industrial recombinant vaccine antigen production. Vaccines (Basel) 9:1504
3. Yamaguchi M, VanderLinden R, Weissmann F, Qiao R, Dube P, Brown NG, Haselbach D, Zhang W, Sidhu SS, Peters JM, Stark H, Schulman BA (2016) Cryo-EM of mitotic checkpoint complex-bound APC/C reveals reciprocal and conformational regulation of ubiquitin ligation. Mol Cell 63:593–607
4. Tian W, Li B, Warrington R, Tomchick DR, Yu H, Luo X (2012) Structural analysis of human Cdc20 supports multisite degron recognition by APC/C. Proc Natl Acad Sci USA 109:18419–18424
5. Zhang Z, Yang J, Barford D (2016) Recombinant expression and reconstitution of multiprotein complexes by the USER cloning method in the insect cell-baculovirus expression system. Methods 95:13–25
6. Lara-Gonzalez P, Pines J, Desai A (2021) Spindle assembly checkpoint activation and silencing at kinetochores. Semin Cell Dev Biol 117: 86–98
7. Van Regenmortal MHV, Fauquet CM, Bishop DHL, Cartsens EB, Estes MK, Lemon SM, Maniloff J, Mayo MA, McGeoch DJ, Pringle CR, Wickner RB (2000) Virus taxonomy: classification and nomenclature of viruses. Seventh report of the international committee on taxonomy of viruses. Academic Press, San Diego
8. Chen G, Zhao S, Chen N, Wu X (2022) Molecular mechanism responsible for the hyperexpression of baculovirus polyhedrin. Gene 814:146129
9. Smith GE, Summers MD, Fraser MJ (1983) Production of human beta interferon in insect cells infected with a baculovirus expression vector. Mol Cell Biol 3:2156–2165
10. Pennock GD, Shoemaker C, Miller LK (1984) Strong and regulated expression of Escherichia coli beta-galactosidase in insect cells with a baculovirus vector. Mol Cell Biol 4:399–406
11. Liu F, Wu X, Li L, Liu Z, Wang Z (2013) Use of baculovirus expression system for generation of virus-like particles: successes and challenges. Protein Expr Purif 90:104–116
12. Kitts PA, Possee R (1993) A method for producing recombinant baculovirus expression vectors at high frequency. BioTechniques 14: 810–817
13. Chambers A, Aksular M, Graves L, Irons S, Possee R, King L (2018) Overview of the baculovirus expression system. Curr Protoc Protein Sci 91:541–546
14. Cosma MA, Curtis NL, Pain C, Kriechbaumer V, Bolanos-Garcia VM (2022) Biochemical, biophysical, and functional characterisation of the E3 ubiquitin ligase APC/C

regulator CDC20 from Arabidopsis thaliana. Front Physiol 13:938688

15. Cheung RC, Wong JH, Ng TB (2012) Immobilized metal ion affinity chromatography: a review on its applications. Appl Microbiol Biotechnol 96:1411–1420
16. Bolanos-Garcia VM, Davies OR (2006) Structural analysis and classification of native proteins from E. coli commonly co-purified by immobilised metal affinity chromatography. Biochim Biophys Acta 1760:1304–1313

Chapter 5

Quantitative Live-Cell Imaging to Study Chromatin Segregation and Nuclear Reformation

Anja Scheufen and Daniel Moreno-Andrés

Abstract

Live-cell imaging is a powerful tool for the investigation of different steps of the life and fate of single cells and cell populations. In this chapter, we describe how to perform live-cell imaging in tissue culture cells and the subsequent image analysis to precisely characterize the cytological events occurring during mitotic exit and nuclear reformation.

Key words Live-cell imaging, Image analysis, Mitotic exit, Chromatin decondensation, Nuclear reformation

1 Introduction

Different modalities of live-cell imaging are state of the art to accurately identify the nature of the cell cycle and mitotic defects in model organisms and tissue cultures. More specifically, the quantitative analysis of time-lapse recordings of mitotic cells allows to measure dynamic changes in subcellular compartments to accurately characterize linked molecular processes in different experimental conditions and clinically relevant situations [1–6]. However, most of these efforts focus on studying early mitotic processes, i.e., prophase, prometaphase, and metaphase. In turn, the study of the mechanisms that drive mitotic exit, chromatin decondensation, and enable nuclear reassembly has received much less attention [7, 8]. Yet, events during anaphase, telophase, and early G1 phase also make important contributions to nuclear homeostasis [9–15] and their malfunction affects genomic integrity, causing different pathologies and inflammation [16–18].

Here we describe the protocols that we routinely use for low and high-resolution live-cell imaging of HeLa cells to obtain quantitative measures of time-dependent cell features during mitotic exit, anaphase, telophase, and early G1. Importantly, these

Victor M. Bolanos-Garcia (ed.), *Mitotic Exit: Methods and Protocols*, Methods in Molecular Biology, vol. 2874, https://doi.org/10.1007/978-1-0716-4236-8_5,

protocols allow mitotic-stage specific analysis without the need for drug-dependent synchronization, e.g., nocodazole-mediated mitotic arrest of the cell population and subsequent release. This reduces the complexity of the experimental setup and avoids the out-of-target effects of combined treatments. Furthermore, the protocols are applicable to a wide variety of tissue culture cell lines. We used them for different adherent (HeLa, RPE, U2OS) [13, 14, 19, 20] or non-adherent (32D, TF-1) [6] cell lines after adjusting the seeding conditions, transfection and labeling protocols, and imaging parameters to avoid growth inhibition due to cell contact, phototoxicity, and cell death.

2 Materials

2.1 Preparation of Cells for Imaging

1. HeLa cells stably expressing H2B-mCherry and/or different organelle markers (*see* **Note 1**). Long-term storage possible at −150 °C or liquid nitrogen.
2. Dulbecco's modified Eagle's medium (DMEM). Store at 4 °C.
3. Fetal bovine serum (FBS). Store at −20 °C.
4. Penicillin-streptomycin solution: 10,000 U/mL penicillin and 10 mg streptomycin/mL (*see* **Note 2**). Store at −20 °C.
5. Phosphate-buffered saline (PBS). Store at 4 °C.
6. Opti-MEM I Reduced Serum Medium. Store at 4 °C.
7. Plasmids for transient transfection of fluorescent protein fusions and/or siRNA oligonucleotides for downregulation of gene expression. Store at −20 °C (*see* **Note 1**).
8. Lipofectamine RNAiMax. Store at 4 °C.
9. JetPRIME (Polyplus, Illkirch, France). Store at 4 °C.
10. SiR-Hoechst, Sir-Tubulin (Spirochrome). Store at −20 °C (*see* **Note 1**).
11. Chambered coverslip with 8 or 18 individual wells and high walls, ibidi treat for cell culture, immunofluorescence, and high-end microscopy (*see* **Note 3**).

2.2 Instruments for Imaging

1. Inverted microscope Nikon Ti2-E with fully motorized XYZ stage and perfect focus system (PFS) (*see* **Note 4**).
2. Dual operation module (confocal or fluorescence widefield) Crest X-Light V2 Spinning Disc with Single Pattern–22 mm/40 μm Pinhole optimized for 60× Objective. Large-FOV: 22 × 22 mm (*see* **Note 5**).
3. Objectives: 60× Lambda Oil 1.4NA WD 0.13 mm, 40× Lambda Air 0.95NA WD 0.16–0.25 mm, 20× Lambda Air 0.75NA WD 1 mm, 10× Lambda Air 0. 5NA WD 2 mm.

4. Metal halide 100 W lamp for epifluorescence.
5. LED Light engine Lumencor Spectra X for Crest Spinning Disk (*see* **Note 6**).
6. Illumination/filter set compatibility: DAPI, Hoechst, CFP, GFP, FITC, YFP, TRITC, Cy3, Texas Red, mCherry, Cy5.
7. Zyla 4.2 USB3 SCMOS Camera, >80%QE, 40ffps@16bit, 53 ffps@12bit.
8. OKOLAB UNO-T-H-CO_2 top-stage incubator system with CO_2 mixer and temperature regulation compatible with two 8well u-Slices and two 35 mm u-Dishes and 24 and 96 well plates (*see* **Note 7**).
9. High-performance computing workstation with specific hardware and software (Nikon Elements AR®) to control the microscope, optic path, illumination, and cameras. Importantly, the setup should allow for automatic multiposition recording, as well as an advanced software-based autofocus module (i.e., Nikon JOBs® module) to implement complex pipelines for three-dimensional cell tracking "on-the-fly" to perform high-resolution recordings.

2.3 Hardware and Software for Image Analysis

1. High-performance computing workstation for image analysis which runs on a 64-bit operative system (OS). Ideally, the system is mounted with a multicore processor (> 2.4GHz/30MB), >32 GB of RAM, >500GB SSD for the OS, 1 TB SATA drive for storage, and a dedicated graphic card with >4GB (*see* **Note 8**).
2. Analysis software:
   - Open FiJi/ImageJ 64-bit [21] (https://imagej.net/; https://fiji.sc).
   - CellCognition [3] (https://cellcognition-project.org).
   - LiveCellMiner [14] (https://github.com/stegmaierj/LiveCellMiner).

3 Methods

3.1 Cell Transfection/Treatment

1. Maintain HeLa cells actively growing in log phase (confluency between 40% and 80%) in DMEM supplemented with 10% FBS and appropriate antibiotics at 37 °C and 5% CO_2 in a CO_2 incubator. We suggest passing them at the appropriate density the day before the protocol starts (*see* **Note 1**).
2. Wash the cells twice with PBS for 1 min (*see* **Note 9**).

3. Trypsinize the cells with phenol red-free trypsin–EDTA and resuspend in DMEM 10% FBS with the appropriated antibiotics (*see* **Note 10**).
4. Prepare 8000–12,000 cells in 250 μL of growth medium.

3.1.1 Transfection of Plasmids

1. Plate the cells 8000–12.000 cells in 250 μL of growth medium in each well of an 8 well Ibidi chamber (*see* **Note 11**).
2. Allow the cells to attach and grow for 24 h at 37 °C.
3. Transfect the cells with the suitable fluorescent fusions using jetPRIME. Per well of 8 well Ibibi chamber, dilute 0.5 μg of DNA in 50 μL of jetPRIME buffer. Mix by vortexing 5 s. Add 1 μL of jetPRIME reagent, vortex for 10 s, spin down 10 s. Incubate for 10 min at room temperature (*see* **Note 12**).
4. Add 50 μL transfection mix drop wise in each of the 8 well from the Ibidi chamber (Subheading 3.1) and return the plates into the incubator.
5. After 4 h, replace transfection medium with DMEM 10% FBS without selection antibiotics and return the plates to the incubator (*see* **Note 13**).
6. Imaging can start when the expression the fusion proteins is detectable with illumination settings below the previously determined threshold of toxicity. This is typically the case between 10 and 24 h post-transfection.

3.1.2 Transfection of siRNA

1. Transfect the cells with the suitable small interference RNAs (siRNAs) using Lipofectamine RNAiMAX at a final concentration of 20 nM. For a single well of 8 well Ibibi chamber, dilute in a tube 0.6 μmol of siRNA in 25 mL of Opti-MEM buffer and mix by pipetting. In a separate tube dilute 0.3 μL of Lipofectamine RNAiMAX reagent in 25 μL of Opti-MEM buffer, mix by pipetting, and incubate 5 min at RT. Mix the contend of both tubes by pipetting and incubate 20 min at RT (*see* **Note 12**).
2. Add 50 μL of transfection mix in each well of an 8 well IBIDI chamber.
3. Slowly add 8000–12,000 cells in 250 μL growth medium by dropping them on the transfection mix (*see* **Note 11**).
4. Imaging acquisition can start typically between 24 and 48 h post-transfection (*see* **Note 14**).

3.2 Imaging

1. Warm up the top-stage incubator system and allow for gas equilibration for 20 min.
2. Incubate the chamber with the cells in the top-stage incubator for 1 h before start imaging to avoid focal drift due to thermal fluctuations.

3.2.1 Live Cell Imaging at Low Resolution for High Content Analysis of Early and Late Mitosis and G1

1. Select the light path in fluorescent widefield mode and 10× or 20× air objectives.
2. Select the multiposition list with the fields of view to be recorded in each well of the transfected/treated chamber plane using the ocular and epifluorescence illumination. Calibrate the multiposition list in XY and Z using the PFS module. Typically, no more than four positions per well should be chosen each separated by more than 200 μm (*see* **Note 15**).
3. Set up the illumination conditions according to labeling, expression of fluorescent fusion proteins, and the phototoxic tolerance of the cells.
4. Set up time-lapse conditions, i.e., frame rate and duration of the record. Typically, an image per position every 3 min during 24–72 h post-transfection is possible for HeLa and RPE cells without phytotoxicity effects (*see* **Note 16**).

3.2.2 Live Cell Imaging at High-Resolution Tracking of Subcellular Structures During Late Mitosis and G1

1. Select the light path in confocal mode and, depending on the required resolution, the 40× 0.95NA air or the 60× 1.4NA oil objective.
2. Place the well with the cells to be imaged on the objective and find the focal plane using the ocular and epifluorescence illumination.
3. Employ the software-based autofocus pipeline in the Nikon JOBs® module. The pipeline uses very low intensity Z-sweep with the chosen illumination settings to find the best-in-focus optical section for a region of interest for the respective filter setting. For our analyses, the pipeline is typically adapted to determine the center of the chromatin mass in Z. Then, it applies the chosen illumination conditions according to cell labeling, expression of fluorescent fusion proteins, and the phototoxic tolerance of the cells for imaging at optimal resolution (*see* **Note 17**).
4. Search for a mitotic cell, ideally in metaphase. With the 40× Air objective several cells located in the same or different wells can be tracked. As for the 60× immersion oil objective, multiposition is allowed in a short range of XY displacement (less than 100 μm). In these conditions, with a sufficient large drop of immersion oil, three closely located positions can be tracked for up to 2 h. Save the table of coordinates as XML file.
5. Load the table of coordinates in the software-based autofocus pipeline and start time-lapse recording at the adequate frame rate. Recording time is typically 1–3 h for anaphase, telophase/G1 events.

6. Stop recording and save the files manually. Ideally, several (4–8) control and treated cell can be recorded within in 10 h of microscope use.

3.3 Image Analysis

1. Use the acquisition proprietary software (Nikon Elements AR®) to convert time-lapse (multiposition) image files into a free open format, i.e., TIFF, OME-TIFF, or similar. For time-lapse multiposition files, it is advisable to split the original file according to the different positions identified by the acquisition order. As high content live-cell imaging might have many positions and hundreds of GBs, it is advisable to transform the original file into a file structure where the images of each position are contained in a folder. The images of each position should be organized sequentially by a standardized naming scheme readable by the analysis software. For example, P001_t0001_z1_c1, where "P" indicates the position on the well/plate, "t" indicates the ordinal frame, "z" indicates the slice in the stack, and c indicates the filter channel.

3.3.1 Automatic Analysis of Dwell Times in Anaphase and Telophase (CellCognition)

1. Use the CellCognition software [3] to measure the length of different phases of mitosis, including early and late anaphase, telophase/G1 based on chromatin segmentation in hundreds to thousands for cells as in [12, 22] (*see* **Note 18**).
2. Load the image file structure from Subheading 3.3.1 in CellCognition.
3. Train the machine-learning based classifier to identity the different mitotic phases based in chromatin morphology.
4. Eventually, additional classifiers can be trained for additional channels to determine dwell times for cellular events like the changes in spindle morphology [22], Golgi reassembly or nuclear envelope (NE) integrity based on the nuclear import of transport substrates, e.g., the importin β binding domain of importin α fused with eGFP as a nuclear import substrate [3].
5. Run the software to segment and track individual cells, synchronize individual cell trajectories according to a given mitotic event, in this case metaphase to anaphase transition.
6. Extract the desired dwell times from the CellCognition output.
7. During the process of analysis, CellCognition can generate galleries of single mitotic cells. If the nature of the event under analysis does not allow automatic analysis, but the resolution of the record matches the nature of the event, visual inspection and manual assessment can be done following the instruction in point Subheading 3.3.3. This allows, for example, to quantify chromatin segregation errors and micronuclei as in [6].

3.3.2 Automatic Analysis of Dynamic Features of Labeled Cell Structures (LiveCellMiner)

1. Use the LiveCellMiner software [14] to measure dynamic phenotypes of labeled cell compartments during mitosis. It is specially developed to quantify time-dependent cell features during mitotic exit, anaphase, telophase/G1 (*see* **Note 19**).
2. Load the image file structure from Subheading 3.3.1 in LiveCellMiner.
3. When employing LiveCellMiner, the program will automatically recognize, segment, track, and align single mitotic cell trajectories in anaphase onset (Fig. 1a) to extract and present quantitate time-resolved image features (Fig. 1b).
4. The LiveCellMiner simple graphical user interface is designed to visually check the automatic mitotic alignment (Fig. 1a) in case the default classical algorithms do not correctly recognize trajectories or transitions. This graphical interface allows man-machine feedback to generate the manual annotation of few cell trajectories which will be used by a neural network for error correction in alignment and wrongly selected non-mitotic cell trajectories.
5. The dynamic features of chromatin and additional fluorescent protein fusions or chemical live-cell labelings can be quantitated to measure changes spindle dynamics as well as nuclear envelope recruitment, integrity or transport competence or similar features (Fig. 1a, b) [14].
6. LiveCellMiner can generate galleries of single mitotic cells. If the nature of the event under analysis does not allow automatic analysis, visual inspection, and manual assessment is also possible following the instruction in point Subheading 3.3.3 (Fig. 1c, d).

3.3.3 Analysis of Labeled Cell Structures at High Resolution (Fiji)

1. Use FiJi (ImageJ) [21] to extract and assemble multichannel image galleries or time-lapse stacks from cell trajectories during mitotic exit from original or converted files (Subheading 3.3.1) (Fig. 2), as well as from galleries from CellCognition or LiveCellMiner (Fig. 1c, d).
2. Quantitative measurement of signal intensities can be extracted using ROIs of the chromatin and cytoplasm sampled using the FiJi "segmented line" tool and the ROI manager. The values can be exported as .csv or excel tables for further processing. For example, the mean signal intensity at the chromatin border can be determined after subtraction of the cytoplasmic background. (Fig. 2a, b) (*see* **Note 20**).
3. By visual inspection and manual assessment, the average delay on recruitment or re-appearance of subcellular structures can be determined with respect to the last metaphase frame/anaphase onset. For example, the time of first contact of NE membrane structures on the surface of the decondensing

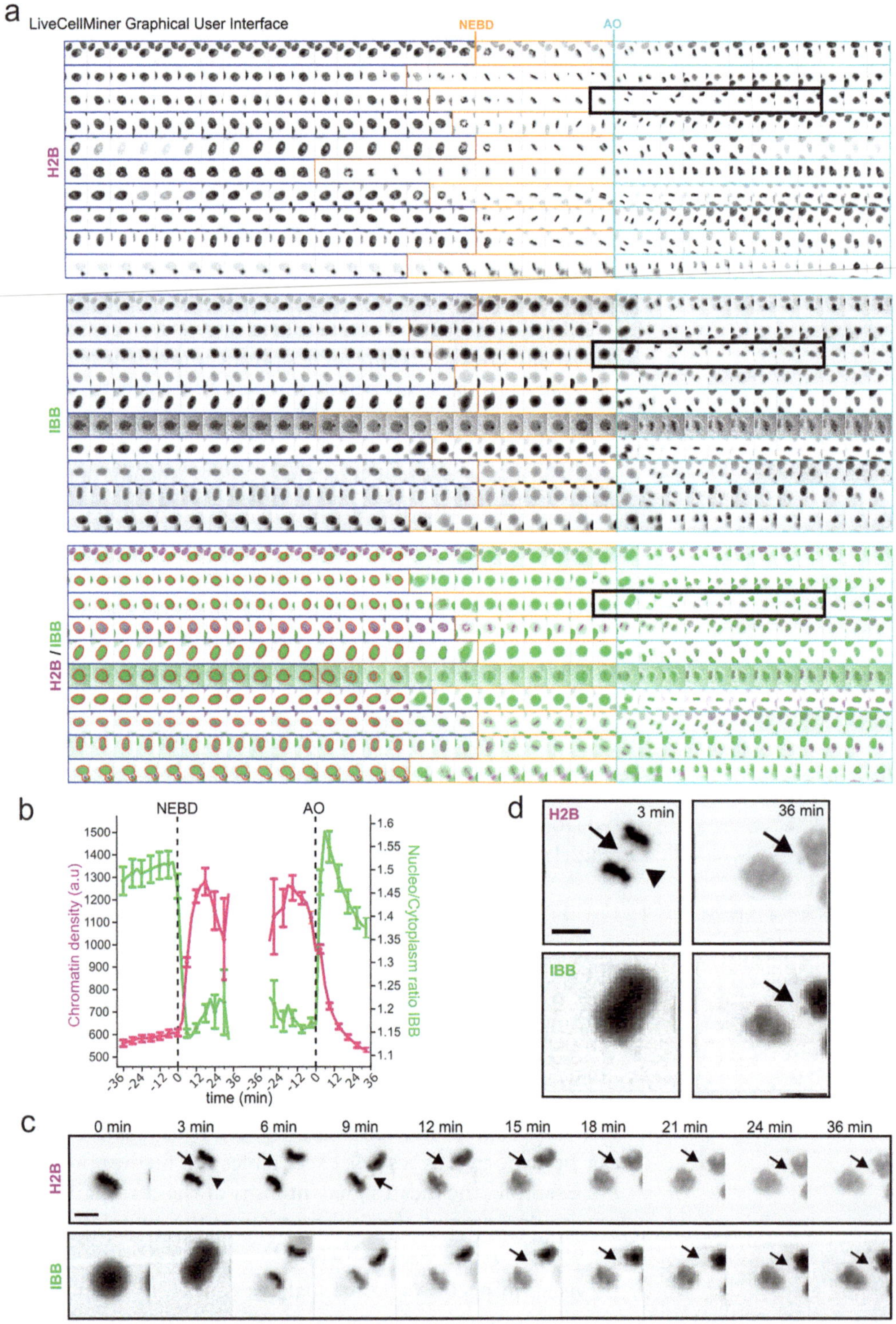

Fig. 1 Automatic and manual analysis of dynamic features of labeled late-mitotic cell structures with LiveCellMiner. (**a**) Graphical user interface of LiveCellMiner showing automatically extracted mitotic events, aligned at the anaphase onset. For each cell, the period before mitotic entry at the end of G2 is boxed in dark blue, the period from nuclear envelope breakdown (NEBD) to anaphase onset (AO) is boxed in orange, and the subsequent mitotic exit and nuclear reformation period is boxed light blue. The upper panel (H2B) shows the

chromatin, cleavage furrow formation or the closure of the NE can be annotated by an expert user (Fig. 2c), to quantify timing under different experimental conditions as in [13].

4 Notes

1. The fluorescent markers must be adequate for the process under study. Usually, chromatin labeling is necessary to determine the mitotic phases and to follow the fate of the newly formed nucleus. This labeling is commonly achieved by fluorescent fusions of H2B or live-compatible chemical labeling such SiR-Hoechst [24]. This chemical labeling needs careful optimization for each experimental setup to avoid toxicity [25, 26]. Fluorescent fusion marker proteins can visualize additional cell structures. For example, inner nuclear envelope proteins such as Lap2β can be used to visualize nuclear envelope recruitment and closure. Fluorescent fusions with tubulin or actin can be used to visualize cytoskeleton dynamics. Alternatively, chemical labeling for time-lapse microscopy of cytoskeletal structures such SiR-Tubulin or Sir-Actin are available. The stable expression of fluorescent fusion proteins is of advantage since they show lower cell-to-cell variability and are usually near to native levels of expression. However, transient transfections can be used, if overexpression is not toxic, but the results need to be interpreted with caution.
2. If the cell lines express additional fluorescent markers, additional antibiotics might be necessary.
3. Other chambers for high-end microscopy, i.e., Labteck, are also suitable.
4. Any inverted microscope with an XYZ-motorized stage enabling a constant focus offset is also suitable.

Fig. 1 (continued) images of the chromatin, labeled by a H2B-mCherry fusion. The middle panel (IBB) shows the images of the importin β binding domain of importin α fused with eGFP as a nuclear import substrate (kind gift from Daniel Gerlich; [9]). The lower panel shows the merge of H2B and IBB signal as well as the segmented region of interest (ROI) of the chromatin in red. These mitotic trajectories are obtained from hundreds (400) of mitotic events recorded every 3 min in a population of unsynchronized HeLa cells cultured in DMEM supplemented with 10% FBS and pen/strep for 24 h. (**b**) Plot showing the temporal change of chromatin density based on the mean H2B-mCherry intensity (arbitrary units AU) and the import competence of the nuclear envelop based on the nuclear/cytoplasm ratio of the IBB-eGFP. Mean and SD is shown for 400 mitotic events aligned to the phase transition at NEBD and AO. (**c**) Gallery of a mitotic cell with chromatin segregation errors such as lagging chromosomes (arrows) that become micronuclei. Scale bar 10 μm. (**d**) Detailed visual inspection of (**c**) allows the quantitation of chromatin segregation errors and nuclear envelope competence for nuclear transport into micronuclei. The arrow points to a lagging chromosome which will develop into a micronucleus, and the arrowhead indicates a thin mitotic bridge. Scale bar 10 μm

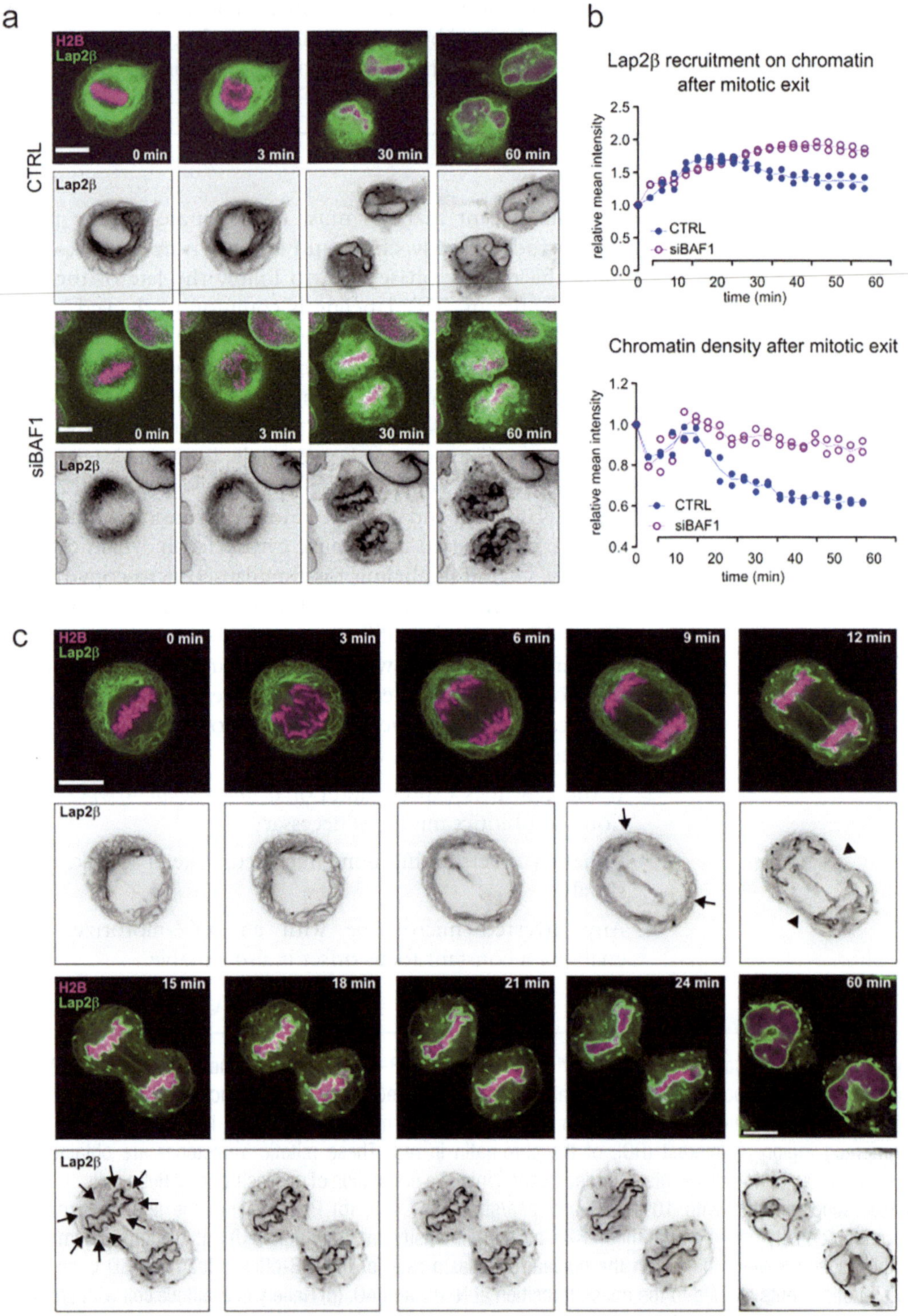

Fig. 2 Analysis of labeled cell structures at high resolution. (**a**) Gallery at high resolution (40× NA0.95 objective in confocal mode) of HeLa cells during mitotic exit and nuclear reformation, expressing H2B-mCherry and eGFP-Lap2β (kind gift from Daniel Gerlich; [23]) transfected with 20 nM siRNA oligos against BANF1 or control and record 72 h after transfection. The cells were subjected to a multiposition time lapse (frame rate 3 min) with five planes in Z spanning the entire vertical displacement that the cells follow from the relatively loose

5. The dual operation confocal or fluorescence widefield is not mandatory and different microscopy systems could be used depending on the nature of the events to track and the resolution needed.
6. Laser light sources or fluorescent lamps depend on the nature of the events to track and the resolution needed.
7. In the case of short-term live-cell imaging (< 4 h), a CO_2-independent medium can be used instead of the more expensive climatic top-stage chamber with a CO_2 mixer.
8. For efficient handling and conversion of large image data sets and the use of CellCognition or LiveCellMeiner, we recommend a high-performance computing workstation. Usually, the workstation that controls the microscope might be used for this purpose but will reduce the available microscopy time. Alternatively, computing clusters can be used for the analysis of large data sets. In addition, standard computers with or above 16GB RAM, CPU from Intel® Core™ i5 or AMD Ryzen™ 5, and a dedicated GPU are sufficient for the analysis described in point Subheading 3.3.3.
9. Alternatively, tightly adherent cells could also be washed with the trypsin solution instead of PBS to seed up cell detachment and separation.
10. The time of incubation and volume of trypsin must be adapted to the cell line and plate format. We recommend using a volume of trypsin sufficient to cover the layer of cells, but negligible compared to the volume of DMEM used to resuspend them. In this way, additional washing and centrifugation steps that could affect cellular fitness are avoided.
11. The number of cells seeded must be optimized for each cell line depending on the treatment and the time post-seeding when they will be imaged to record cells growing in log-phase.

Fig. 2 (continued) attachment at beginning of mitotic exit until they completely adhere and expand in G1 (*see* **Note 15**). The best in-focus plane of each time point was selected for analysis. Scale bar 10 μm. (**b**) Plots showing abnormal initial nuclear envelope chromatin contact and delayed chromatin condensation in a BANF1 downregulated cell (magenta circles and line) as compared to a control cell (blue circles and line) [23]. FiJi "segmented line" tool and the ROI-manager were used to measure the signal intensities of chromatin and Lap2β in the nuclear compartment at each time point during 1 h after metaphase to anaphase transition were measured using the FiJi "segmented line" tool and the ROI-manager. The values were exported as excel tables and after background subtraction and normalization to last metaphase frame plotted using GraphPad. (**c**) Gallery at high resolution (60× NA1.4 objective in confocal mode with software-based autofocus Nikon JOBs®) of HeLa cells during mitotic exit and nuclear reformation, expressing H2B-mCherry and eGFP-Lap2β. The average delay of the first contact of nuclear envelope (arrows) with the chromatin, cleavage furrow (arrow heads) formation, and full enclosure of nuclear envelope with respect to the last metaphase/anaphase frame can be quantified by visual inspection of several cells in different experimental conditions. Scale bar 10 μm

12. The amounts of the transfection DNA/JetPrime-reagent ratios must be optimized for each cell line and construct. Initial analysis can be carried out using a fluorescence microscope to assess the time after transfection when the cells have detectable fluorescent signals and measure transfection efficiency, cell viability, and health.
13. The incubation time with the transfection mix must be optimized for each cell line and construct. Medium change might vary from 1 to 24 h or not be needed at all depending on the combined toxicity of the transfection process and the expression of the protein fusion in a particular cell type. In this step, if necessary to increase the signal-to-noise ratio, a specific, phenol red-free growth medium, i.e., FluoroBrite™ DMEM, can be used.
14. To estimate the time of starting microcopy recording, it is recommended to determine by Western blotting at which time point after siRNA transfection the expression of the target protein is sufficiently reduced.
15. The number of total positions will be limited by the image frequency (e.g., one image per position every 3 min), the number of channels, and the speed at which the microscope hardware can accomplish the changes in the light path and the XYZ movements to focus each position. For example, the system used here, in widefield mode, can record 32 positions in two channels every 2 min and 40–50 s. This allows recording 32 dual channel positions every 3 min with a time buffer for eventual automatic adjustment of the perfect focus system for hours or days.
16. Careful optimization of illumination and time-lapse conditions according to the expression of fluorescent fusions and/or the intensity and nature of the chemical labeling are mandatory to avoid phototoxicity.
17. In absence of software and/or hardware-based autofocus, similar outputs can be achieved by the acquisition of several planes in Z per position at each time point, spanning the entire vertical displacement of the cell from the relatively loose attachment during early mitotic exit until it completely adheres and expands in G1 (Fig. 2c). Careful optimization of illumination and time-lapse conditions is needed to avoid phototoxicity.
18. CellCognition has a standalone user interface, which is not completely intuitive, and thus needs dedicated training. For further information and instructions see [3] and visit (https://cellcognition-project.org).
19. LiveCellMiner is a simple module of the ScixMiner extension for MATLAB. The user does not require programing skills. It is easy to learn yet requires dedicated training. For further

information and instructions, see https://github.com/stegmaierj/LiveCellMiner [14].

20. FiJi/ImageJ is a standalone multipurpose image analysis program. The user does not require programing skills. It is easy to learn yet requires dedicated training. For further information and instructions, see https://imagej.net/ and https://fiji.sc [21].

Acknowledgment

We thank Dr. Mohamed Hamed for the seeding of the cells presented in Fig. 1.

References

1. Neumann B et al (2006) High-throughput RNAi screening by time-lapse imaging of live human cells. Nat Methods 3(5):385–390. https://doi.org/10.1038/nmeth876
2. Neumann B et al (2010) Phenotypic profiling of the human genome by time-lapse microscopy reveals cell division genes. Nature 464(7289):721–727. https://doi.org/10.1038/nature08869
3. Held M et al (2010) CellCognition: time-resolved phenotype annotation in high-throughput live cell imaging. Nat Methods 7(9):747–754. https://doi.org/10.1038/nmeth.1486
4. Failmezger H, Frohlich H, Tresch A (2013) Unsupervised automated high throughput phenotyping of RNAi time-lapse movies. BMC Bioinformatics 14:292. https://doi.org/10.1186/1471-2105-14-292
5. Yokoyama H et al (2019) Chromosome alignment maintenance requires the MAP RECQL4, mutated in the Rothmund-Thomson syndrome. Life Sci Alliance 2(1). https://doi.org/10.26508/lsa.201800120
6. Holl K et al (2024) Calreticulin and JAK2V617F driver mutations induce distinct mitotic defects in myeloproliferative neoplasms. Sci Rep 14(1):2810. https://doi.org/10.1038/s41598-024-53240-8
7. Vagnarelli P (2021) Back to the new beginning: mitotic exit in space and time. Semin Cell Dev Biol 117:140–148. https://doi.org/10.1016/j.semcdb.2021.03.010
8. Moreno-Andrés D, Holl K, Antonin W (2023) The second half of mitosis and its implications in cancer biology. Semin Cancer Biol 88:1–17. https://doi.org/10.1016/j.semcancer.2022.11.013
9. Schmitz MH et al (2010) Live-cell imaging RNAi screen identifies PP2A-B55alpha and importin-beta1 as key mitotic exit regulators in human cells. Nat Cell Biol 12(9):886–893. https://doi.org/10.1038/ncb2092
10. Schneider MWG et al (2022) A mitotic chromatin phase transition prevents perforation by microtubules. Nature. https://doi.org/10.1038/s41586-022-05027-y
11. Cuylen-Haering S et al (2020) Chromosome clustering by Ki-67 excludes cytoplasm during nuclear assembly. Nature 587(7833):285–290. https://doi.org/10.1038/s41586-020-2672-3
12. Moreno-Andres D et al (2020) VPS72/YL1-mediated H2A.Z deposition is required for nuclear reassembly after mitosis. Cells 9(7). https://doi.org/10.3390/cells9071702
13. Holzer G et al (2021) The nucleoporin Nup50 activates the ran guanine nucleotide exchange factor RCC1 to promote NPC assembly at the end of mitosis. EMBO J 40(23):e108788. https://doi.org/10.15252/embj.2021108788
14. Moreno-Andres D, Bhattacharyya A, Scheufen A, Stegmaier J (2022) LiveCellMiner: a new tool to analyze mitotic progression. PLoS One 17(7):e0270923. https://doi.org/10.1371/journal.pone.0270923
15. Orr B et al (2021) An anaphase surveillance mechanism prevents micronuclei formation from frequent chromosome segregation errors. Cell Rep 37(6):109783. https://doi.org/10.1016/j.celrep.2021.109783

16. Manchado E et al (2010) Targeting mitotic exit leads to tumor regression in vivo: Modulation by Cdk1, Mastl, and the PP2A/B55alpha, delta phosphatase. Cancer Cell 18(6): 641–654. https://doi.org/10.1016/j.ccr.2010.10.028
17. Zierhut C, Funabiki H (2020) Regulation and consequences of cGAS activation by self-DNA. Trends Cell Biol 30(8):594–605. https://doi.org/10.1016/j.tcb.2020.05.006
18. Li T et al (2021) Phosphorylation and chromatin tethering prevent cGAS activation during mitosis. Science 371(6535). https://doi.org/10.1126/science.abc5386
19. Cai Y et al (2018) Experimental and computational framework for a dynamic protein atlas of human cell division. Nature 561(7723): 411–415. https://doi.org/10.1038/s41586-018-0518-z
20. Wang P et al (2021) Loss of haspin suppresses cancer cell proliferation by interfering with cell cycle progression at multiple stages. FASEB J 35(10):e21923. https://doi.org/10.1096/fj.202100099R
21. Schindelin J et al (2012) Fiji: an open-source platform for biological-image analysis. Nat Methods 9(7):676–682. https://doi.org/10.1038/nmeth.2019
22. Schooley A et al (2015) The lysine demethylase LSD1 is required for nuclear envelope formation at the end of mitosis. J Cell Sci 128(18): 3466–3477. https://doi.org/10.1242/jcs.173013
23. Samwer M et al (2017) DNA cross-bridging shapes a single nucleus from a set of mitotic chromosomes. Cell 170(5):956–972. https://doi.org/10.1016/j.cell.2017.07.038
24. Lukinavicius G et al (2015) SiR-Hoechst is a far-red DNA stain for live-cell nanoscopy. Nat Commun 6:8497. https://doi.org/10.1038/ncomms9497
25. Sen O, Saurin AT, Higgins JMG (2018) The live cell DNA stain SiR-Hoechst induces DNA damage responses and impairs cell cycle progression. Sci Rep 8(1):7898. https://doi.org/10.1038/s41598-018-26307-6
26. Rajendraprasad G, Rodriguez-Calado S, Barisic M (2023) SiR-DNA/SiR-Hoechst-induced chromosome entanglement generates severe anaphase bridges and DNA damage. Life Sci Alliance 6(12). https://doi.org/10.26508/lsa.202302260

Chapter 6

Determination of Anaphase Duration by Time-Lapse Microscopy in Budding Yeast

Mariam Huda and Ayse Koca Caydasi

Abstract

Time-lapse microscopy is a valuable, widely employed technique for investigating cell cycle dynamics. This chapter will guide you in using this tool to determine the duration of anaphase as a measure of mitotic exit kinetics in budding yeast cells. The methodology explained here is based on monitoring the mitotic spindle and provides a comprehensive guide covering all aspects, from sample preparation and microscopy to data analysis.

Key words Mitotic exit, Budding yeast, *Saccharomyces cerevisiae*, Mitotic spindle, Anaphase, Spindle elongation, Anaphase onset, Spindle breakdown, Time-lapse microscopy

1 Introduction

The transition from mitosis into a new G1, also known as mitotic exit, is tightly regulated to prevent premature cell division before chromosomes are accurately segregated [1–3]. Inactivation of mitotic cyclin-dependent kinase (CDK) complexes and dephosphorylation of CDK targets trigger mitotic exit. Mitotic cyclin-CDK inactivation requires the degradation of mitotic cyclins, which is mediated by an E3-Ubiqutin ligase named the Anaphase Promoting Complex (APC) [4]. In association with its activator subunit Cdc20, APC promotes the degradation of Securin (Pds1) and S- and M-phase cyclins to drive the onset of anaphase and the inactivation of CDK [5]. APC-Cdc20 is activated at the metaphase-to-anaphase transition after silencing of the Spindle Assembly Checkpoint [6], thus restricting mitotic exit to a time point after sister chromatid separation.

In budding yeast *Saccharomyces cerevisiae*, mitotic cyclin degradation takes place in two sequential waves. APC-Cdc20 triggers the first wave, and the second wave is triggered by APC in association with its activator subunit Cdh1 [7–9]. APC-Cdh1 is essential for

Victor M. Bolanos-Garcia (ed.), *Mitotic Exit: Methods and Protocols*, Methods in Molecular Biology, vol. 2874, https://doi.org/10.1007/978-1-0716-4236-8_6,

the full degradation of mitotic cyclins and, thus, for mitotic exit. The conserved phosphatase Cdc14 allows the formation of APC-Cdh1 through the dephosphorylation of Cdh1 [10]. In addition to APC-Cdh1 activation, Cdc14 promotes mitotic exit by upregulating the levels of CDK inhibitor Sic1, which in turn inactivates CDK [10]. Cdc14 is kept sequestered in the nucleolus until anaphase onset [11]. At the metaphase-to-anaphase transition, through activation of the CdcFourteen Early Anaphase Release (FEAR) pathway, Cdc14 is partially released from the nucleolus to the nucleoplasm [12]. FEAR release of Cdc14 is critical for anaphase-related tasks, including anaphase spindle stabilization and elongation, ribosomal RNA segregation, as well as setting the stage for Mitotic Exit Network (MEN) activation [13–19]. The MEN triggers mitotic exit by promoting the full release of the phosphatase Cdc14, which is essential for downregulating mitotic CDK activity [20, 21]. MEN activity is negatively regulated by the Spindle Position Checkpoint (SPOC), a budding yeast mitotic checkpoint that assures alignment of the mitotic spindle in the mother-to-daughter direction before mitotic exit [22]. Through all these mechanisms, activation of the MEN is restricted to mid-to-late anaphase and only after proper alignment of the mitotic spindle in the mother-to-daughter direction.

Following MEN activation, mitotic CDK is downregulated, the mitotic spindle is disassembled, and cytokinesis is initiated [23]. Thus, the time spent from the onset of anaphase until spindle breakdown, which we refer to as the anaphase duration, can be used as an indicator of MEN activity and, thus, an indicator of mitotic exit timing. Mutants with reduced MEN activity have longer anaphase duration [24]. Anaphase duration can be studied in cells with mispositioned spindles to understand functionality of the SPOC as well. In this case, it will be necessary to induce spindle mispositioning in cells by deleting genes involved in spindle positioning (i.e., *KAR9*) [25]. In cells with misaligned spindles, mutants deficient in SPOC have shorter anaphase duration than cells with a functional SPOC [24, 26–29].

Time-lapse microscopy is a powerful tool used to study the cell cycle dynamics. In our lab, we utilize time-lapse microscopy to calculate the anaphase duration of dividing yeast cells. For this, we monitor the mitotic spindle in cells carrying an additional copy α-tubulin (*TUB1*) N-terminally tagged with a fluorophore (i.e., GFP) expressed under the *TUB1* promoter [30, 31]. Time-lapse microscopy allows live imaging of single cells. No synchronization is required, which may otherwise impact normal cell cycle progression. Furthermore, as anaphase is short (~18–20 min) considering the duration of an entire cell cycle (~90 min), the high time resolution of time-lapse microscopy provides a more sensitive platform to allow detection of differences in the anaphase duration. In this chapter, we describe the routine protocol we use for measuring anaphase duration, including tubulin tagging, sample preparation, time-lapse microscopy, and data analysis.

2 Materials

2.1 Growth Medium and Plates

1. 10× SC-His-Trp-Leu-Ura dropout stock: 20 g L-valine, 20 g L-threonine, 20 g L-isoleucine, 5 g adenine hemisulfate salt, 20 g L-cysteine hydrochloride monohydrate, 20 g L-aspartic acid, 20 g L-alanine, 20 g L-arginine monohydrochloride, 20 g L-lysine monohydrochloride, 20 g L-asparagine monohydrate, 2 g 4-aminobenzoic acid, 20 g L-glutamine, 20 g L-glutamic acid potassium salt monohydrate, 20 g glycine hydrochloride, 20 g Myo-inositol, 20 g L-methionine, 20 g L-proline, 20 g L-serine, 20 g L-tyrosine, 20 g L-phenylalanine. Add the above components to a screw cap glass bottle and mix by rotating the bottle for 24 h at room temperature on a roller mixer, then store at 4 °C (*see* **Note 1**).
2. SC-complete dropout mix: To 18.35 g of the 10× SC-His-Trp-Leu-Ura dropout stock prepared in 1, add the following: 1 g L-tryptophan, 1 g L-histidine monohydrochloride, 1 g uracil, and 2 g L-leucine. Mix in a screw cap glass bottle by rotating the bottle for 24 h at room temperature on a roller mixer, then store at 4 °C.
3. SC-URA dropout mix: To 18.35 g of the 10× SC-His-Trp-Leu-Ura dropout stock prepared in 1, add the following: 1 g L-tryptophan, 1 g L-histidine monohydrochloride, and 2 g L-leucine. Mix in a screw cap glass bottle by rotating the bottle for 24 h at room temperature on a roller mixer, then store at 4 °C.
4. Filter-sterilized Synthetic Complete (SC) medium: Dissolve 6.8 g yeast nitrogen base without amino acids, 2 g SC-complete dropout mix, 0.1 g adenine hemisulfate salt, and 20 g glucose in distilled water and complete the volume to 1 L. Then, filter sterilize the solution by filtration through a 0.22 μm bottle top filter. Store at room temperature (*see* **Note 2**).
5. YPD agar plates: Dissolve 10 g Bacto yeast extract, 20 g Bacto peptone, and 20 g glucose in distilled water, complete the volume to 1 L and add 20 g Bacto agar. Autoclave the mixture, stir well, and pour 20–25 mL of the autoclaved mixture into sterile plates.
6. SC-URA agar plates: Dissolve 6.7 g yeast nitrogen base without amino acids, 2 g SC-URA dropout mix, and 20 g glucose in distilled water, complete to 500 ml and autoclave (mixture A). In a separate autoclavable container add 20 g bacto agar and

500 mL water, and autoclave (mixture B). Pour mixture A into mixture B and mix them. Pour 20–25 mL of agar into sterile plates (*see* **Note 3**).

2.2 Obtaining Tubulin-Labeled Yeast

1. A yeast integration vector that carries *GFP-TUB1* under the *TUB1* promoter. Here, we describe using pAFS125 (*URA3*-based yeast integration vector that carries *GFP-TUB1* under *TUB1* promoter) [31] to allow integration into the *URA3* locus.
2. Haploid yeast with available *ura3* mutant locus. Here, we use ESM356-1 of S288C origin with *ura3-52* allele [32].
3. LiSorb: 100 mM Lithium acetate (LiOAc), 10 mM Tris–HCl pH 8, 1 mM EDTA pH 8, 1 M Sorbitol. Filter sterilize and keep at room temperature.
4. LiPEG: 100 mM LiOAc, 10 mM Tris–HCl pH 8, 1 mM EDTA pH 8, 40% PEG3350. Filter sterilize and keep at 4 °C (*see* **Note 4**).
5. Denatured salmon sperm DNA: Heat up sheared salmon sperm DNA (2.0 mg/mL) for 5 min at 95 °C and cool down on ice for 5 min. Keep on ice until usage.
6. DMSO (≥99.7%).
7. 1×PBS: 137 mM NaCl, 2.7 mM KCl, 10 mM Na_2HPO_4, 1.76 mM KH_2PO_4. pH 7.2–7.4, autoclave.

2.3 Sample Immobilization

1. ConA solution: Dissolve 6% (w/v) Concanavalin A (*Canavalia ensiformis* Jack Bean, type IV Sigma C2010-G) in distilled and sterilized water. Make aliquots of 100–200 μL. Store at −20 °C (*see* **Note 5**).
2. Glass-bottom dishes suitable for live cell imaging (e.g., WVR 10810–054 Matsunami or any with a glass well of glass thickness of 0.16–0.19 mm).

2.4 Microscope Setup

The microscope we use for the method described here is a Carl Zeiss Axio Observer 7 motorized inverted epifluorescence microscope. However, any fluorescence microscope with similar properties would suit the method:

1. Motorized inverted microscope (Axio Observer 7 with motorized filter target, motorized table scanning stage and control unit, and anti-vibration unit, Carl Zeiss).
2. CMOS camera (Axiocam 702 Monochrome, Carl Zeiss).
3. Fluorescent illumination source and appropriate excitation/emission filters for visualization of GFP (Colibri 7 LED light source, Filter set 44, Carl Zeiss).

4. 63× or 100× objective suitable for fluorescence imaging (We use Carl Zeiss Plan-Apochromat 63×/1.40 Oil DIC M27, FWD = 0.19 mm).
5. Low or no fluorescence immersion oil (Immersol 518 F, Carl Zeiss, Nikon Type F).
6. Microscope heating unit and incubation chamber (XL incubation unit, Carl Zeiss).
7. Image acquisition software (Zen pro, Carl Zeiss).

2.5 Data Analysis

1. Software for quantification of fluorescence intensity (e.g., Fiji/ImageJ, NIH, Bethesda, Maryland, United States).
2. Software for plotting the data and statistical analysis (e.g., GraphPad Prism, La Jolla, California, United States).

3 Time-Lapsed Microscopy

3.1 Obtaining Tubulin-Labeled Yeast Cells

To observe microtubules in yeast, we integrate into the yeast genome an additional copy of α-tubulin (*TUB1*) N-terminally tagged with a fluorophore (i.e., mCherry, GFP) expressed under *TUB1* promoter. Several tagged *TUB1* constructs are available in yeast integration vectors, allowing integration in different auxotrophic loci [30, 31]. The following method describes generation of haploid yeast strains that carry an additional copy of *GFP-TUB1* under the *TUB1* promoter at the *URA3* locus using pAFS125 [31] as the vector. We also describe the yeast transformation method we routinely use in our lab.

1. Linearize 5 μg of pAFS125 in 50 μL reaction volume using StuI restriction enzyme (*see* **Note 6**).
2. Obtain a log-phase culture in 50 mL YPAD (*see* **Note** 7).
3. Pellet cells at 3200 rpm for 2 min at room temperature.
4. Discard the supernatant, resuspend cells in 45 mL sterilized deionized water, and centrifuge at 3200 rpm for 2 min at room temperature.
5. Discard the supernatant, resuspend cells in 10–20 mL LiSorb, and centrifuge at 3200 rpm for 2 min at room temperature.
6. Discard the supernatant, centrifuge at 3200 rpm for 2 min at room temperature and aspirate any remaining supernatant.
7. For a pellet coming from a 50 mL log-phase culture of OD_{600}, add 300 μL of LiSorb and 50 μL of denatured salmon sperm DNA (*see* **Note 8**), vortex to mix and make 50 μL aliquots of the competent cells (*see* **Note 9**).

8. Pipette 3 μL of the linearized pAFS125 into competent cells, vortex, and leave on the bench at room temperature for 15 min (*see* **Note 10**).
9. Add 300 μL of LIPEG, vortex well to mix, and leave on the bench at room temperature for 15 min (*see* **Note 4**).
10. Add 35 μL of DMSO, vortex, and incubate in a 42 °C water bath for 15 min (*see* **Notes 11** and **12**).
11. Pellet cells at 3200 rpm for 2 min at room temperature, aspirate the supernatant, and resuspend the pellet in 200 μL autoclaved 1×PBS.
12. Plate out the cell suspension on SC-URA agar plates and incubate at 30 °C.
13. Two days later, transformant colonies should appear. Pick six single transformant colonies and streak them onto fresh SC-URA plates separately to obtain single colonies from each initial transformant colony (*see* **Note 13**). Incubate plates at 30 °C.
14. Two days later, pick one isolated single colony coming from each initial transformant colony and patch them on SC-URA plates. Incubate plates at 30 °C.
15. Next day, using a sterile pipette tip, transfer some cells from each patch onto a microscope slide that contains a drop of 1×PBS. Resuspend cells in the 1×PBS to make a wet mount.
16. Image the cells at the bright field and GFP channels, and look for medium and large budded cells to observe the presence of a mitotic spindle in the GFP channel (*see* **Note 14**).

3.2 Culture Growth for Microscopy

Use aseptic conditions to handle all yeast cultures. This protocol describes the growth conditions for strains growing at 30 °C in a glucose-containing medium. Adjust the growth conditions accordingly when using mutant strains or temperature-sensitive mutants that require another temperature, medium composition, or carbon source [33].

1. Two days before your time-lapse experiment, obtain a log-phase culture of *GFP-TUB1* containing cells in filter-sterilized SC-complete media (*see* **Notes 1–2** and 7).
2. Keep your culture in log-phase (OD_{600} < 1.0) in filter-sterilized SC-complete media for about 15 doublings until the time of the experiment (*see* **Note 15**).
3. Use this log-phase culture for the specimen preparation (*see* **Note 16**).

3.3 Specimen Preparation

This method describes the immobilization of yeast cells for live cell imaging. Using this preparation method, cells can be observed for several hours (*see* **Note 17**).

1. Drop 150–200 μL of ConA solution on the glass center of the microwell dish using a micropipette, then spread the ConA over the glass center by tapping the sides of the microwell dish. This amount of ConA works well for a glass well diameter of 14 mm. Close the lid and incubate the dish at room temperature for 10 min.
2. Using a micropipette, remove the ConA solution from the dish (*see* **Note 18**).
3. Wash the dish at least three times using distilled sterile water to remove the unattached ConA from the dish. In each washing step, fill the dish more than halfway with water and aspirate it using a vacuum aspirator (*see* **Note 18**).
4. Add 200 μL of culture (see **Note 16**) to the glass center using a micropipette and close the lid. Incubate at 30 °C for 30 min (*see* **Note 19**).
5. Remove the excess culture from the dish center using a vacuum aspirator.
6. Using preheated (30 °C) filter-sterilized SC medium, wash the dish 3–5 times to remove unattached cells. Follow the washing method described in **Step 3**.
7. Fill the dish at least halfway with the preheated (30 °C) filter-sterilized SC medium, close the lid, and take the dish onto the preheated microscope stage (see below for how to prepare the microscope in advance).

3.3.1 Setting the Stage

The illumination of the specimen with the specific wavelength that excites the fluorophore of interest should be minimized at this stage to avoid photobleaching of the fluorophore during image acquisition (acquisition bleaching).

1. Turn on the microscope heating chamber and adjust the temperature to 30 °C. Let the chamber heat up and reach a thermal equilibrium with the microscope stage and lenses while the sample is prepared (*see* **Notes 20** and **21**).
2. Apply one drop of immersion oil onto the objective lens and place the dish on the objective using an adjustable slide holder.
3. Leave the dish on the microscope stage for 30–40 min before starting image acquisition for the time-lapse experiment (*see* **Notes 22** and **23**). Meanwhile, do the following steps:
4. Focus on cells using bright field.
5. Set the z-stacks for imaging, to span the entire thickness of cells (*see* **Note 24**).

6. Take a few still images in the bright field and GFP channels before the experiment to determine the optimum light intensity and exposure time for the time-lapse experiment (*see* **Note 25**).
7. Select different stage positions for time-lapse imaging (*see* **Note 26**).

3.3.2 Data Acquisition

1. Set the image acquisition software to a time-point acquisition program. Select an appropriate time interval between each time point and an appropriate total imaging duration. In our case, we use a 1-min time interval for a total imaging duration of 60–90 min (*see* **Note 27**).
2. Make sure z-stack mode is selected (*see* **Note 24**).
3. Make sure previously selected stage positions are all focused. If the field of view no longer contains desired cells, replace the stage positions with new ones (*see* **Note 26**).
4. Start time-lapse image acquisition.
5. By monitoring the bright field images occasionally, ensure the cells stay focused until the end of the experiment (*see* **Note 28**).

3.4 Data Analysis

The analysis of the time-lapse requires the usage of image analysis tools. In our set up, Fiji software was used.

1. Open the time-lapse image file in Fiji, selecting one stage position at a time.
2. From the Fiji menu bar, select image → stacks → z-project → sum slices (*see* **Note 29**).
3. In the GFP channel, select a cell that starts with a short spindle (1–2 μm).
4. Using the bright field view, make sure both the mother and daughter compartments of the selected cell are in focus.
5. Start moving through the time points and determine the time point just before the spindle starts fast elongation (*see* **Note 30**). This will be the time point of anaphase onset ($t_{\text{anaphase onset}}$, Fig. 1).
6. Continue moving through the time points until the spindle breaks in the middle and record this time point ($t_{\text{spindle breakdown}}$, Fig. 1).
7. Determine the anaphase duration by subtracting the time point of anaphase onset from the time point of spindle breakdown and multiplying this with the time resolution of your time-lapse. In our case, we had a 1 min time interval between each frame, so it was a 1 min time resolution.

   Anaphase duration $= (t_{\text{spindle breakdown}} - t_{\text{anaphase onset}}) \times (\text{time resolution})$

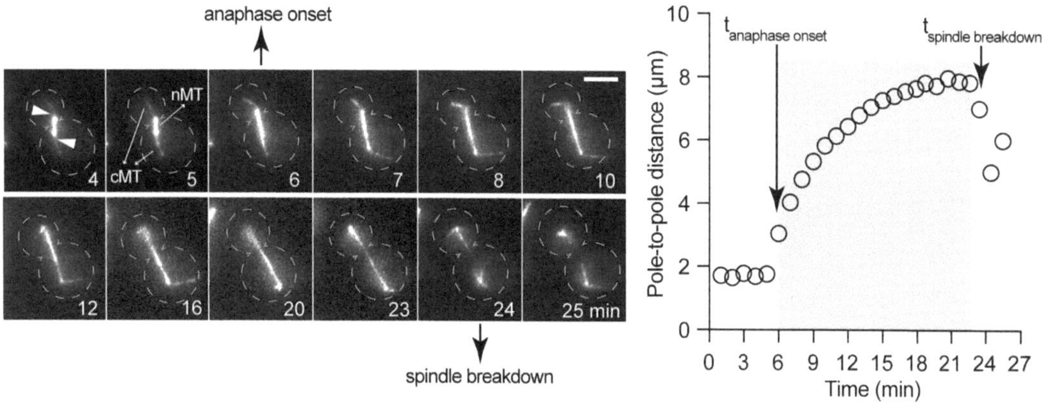

Fig. 1 Monitoring spindle elongation in budding yeast. Selected time points coming from time-lapse microscopy of a *GFP-TUB1* bearing cell. Cell boundaries are outlined with a dashed line. Spindle poles are indicated with arrowheads on the first image. The thick line between poles is nuclear microtubules (nMT) that make the spindle, whereas thin lines projected toward the mother and the daughter cell cortexes are the cytoplasmic microtubules (cMT) that position the spindle. Scale bar is 3 µm. The graph shows how the distance between the two spindle poles changes from metaphase to the end of mitosis. Time points when the spindle started to elongate and breakdown are indicated. The gray shaded area corresponds to the time from anaphase onset until spindle breakdown (anaphase duration)

8. Determine the anaphase duration in as many single cells as possible (*see* **Note 31**).
9. Plot the anaphase duration of individual cells of the population and calculate descriptive statistics (mean, standard deviation).
10. Compare the anaphase duration of different strains by statistical analysis (*see* **Note 32**).

4 Notes

1. Using aminobenzoic acid is crucial, as it reduces phototoxicity during imaging. In addition, we noticed that the viability of the cells during imaging changed when we changed the amino acids' brands.
2. Filter sterilization of the media is critical, as autoclaving results in autofluorescence. Long-term storage of filter-sterilized SC-complete media in a lit room also causes increased background signal, which can be prevented by storage in the dark.
3. To avoid bubble formation while pouring mixture A on mixture B, pour slowly and aim at the sides of the bottle rather than mixture B directly.
4. LiPEG is a highly viscous solution that will take time to pass through the filter. Also, when pipetting, pipette slowly to make sure the desired volume is pipetted.

5. ConA stock solution is reusable if kept at −20 °C. If the solution was left at room temperature for several hours, we advise discarding it.
6. Run 3 μL of the reaction on an agarose gel after restriction digestion to confirm the linearization of the plasmid. The presence of a small amount of undigested integration plasmid along the linearized plasmid would not be a problem, as intact integration plasmids cannot propagate in yeast due to their lack of yeast replication origins.
7. To obtain a log-phase culture, follow the following steps:
   (a) Use aseptic conditions for all yeast work.
   (b) From frozen glycerol stocks, streak yeast cells on YPD agar plates and incubate at 30 °C in a steady incubator until growth can be observed. Once grown, plates can be stored at 4 °C for a month if cells are not cold-sensitive.
   (c) Using a pipette tip, scrape 3–5 colonies from the YPD plate and inoculate in a conical flask containing appropriate media (at least 10 mL). Incubate the flask in a 30 °C shaking incubator (~160 rpm) overnight (~15 h) to obtain a stationary culture the next morning.
   (d) Next morning, measure the optical density of the yeast culture at 600 nm (OD_{600}). For an overnight culture, the OD_{600} is usually expected to be between 5 and 10 for wild-type cells grown in glucose-containing rich media. In our setup, an OD_{600} value of 1 corresponds to 2.2×10^7 yeast cells per ml of culture, and the number of cells/ml correlates with the OD_{600} linearly in an OD_{600} range of 0.1–1. If the OD_{600} readout is higher than 1, make an appropriate dilution of the culture to have an OD_{600} readout within the range of 0.1–1 and multiply with the dilution factor to calculate the actual OD_{600} of your culture. In our hands, 1:10 dilution works for measuring the OD_{600} of overnight cultures.
   (e) In a new conical flask containing fresh growth medium, add an appropriate amount of your culture to obtain an OD_{600} of 0.15. To obtain a log-phase culture, incubate the flask in a 30 °C shaking incubator (~160 rpm) until the OD_{600} of the culture reaches 0.8.
8. The amount of denatured salmon sperm DNA and LiSorb can be scaled up or down depending on the volume and OD_{600} of the culture pelleted. The indicated amount corresponds to the cells pelleted from a 50 mL culture of an OD_{600} value of 1.
9. At this stage, competent cells can be frozen and kept at a −80 °C freezer for further use. Once removed from the

−80 °C freezer, frozen competent cells should be thawed at room temperature. Once thawed, do not freeze cells again.

10. Always include a "no DNA control" for your transformation as a control for contamination.
11. Using a water bath for heating ensures rapid heat transfer.
12. If you are using temperature-sensitive mutants, decrease the incubation time to 12 min.
13. This step ensures that the cells you will work with come from a single genetic origin.
14. Linearized integration plasmid will mostly integrate once into the auxotrophic locus, but it can also integrate multiple times. In this case, multiple integrations will result in a brighter spindle signal, which is more favorable for the time-lapse experiments. To pick the patch that yields the brightest spindle signal, focus on medium-to-large budded cells with long anaphase spindles and compare the spindle brightness of such cells among different patches. Making this comparison using log-phase cells grown in filter-sterilized SC-complete medium may provide better conclusions.
15. Keeping the culture in the log-phase for 15 doublings helps eliminate old cells coming from the stationary culture and also decreases the background fluorescence. To keep the culture in the log-phase for 15 doublings, dilute the initial log-phase culture at appropriate time intervals to always keep the OD_{600} of the culture below 1. When you have to leave the culture overnight, calculate how many times the culture will double overnight and dilute the culture to an initial OD_{600} value to yield an OD_{600} of 0.4–0.8 in the next morning, according to the number of expected doublings.
16. The optimum OD_{600} value to start preparing the specimen for the time-lapse experiment is ~0.5. If your morning culture's OD_{600} is higher, dilute it to ~0.5 before specimen preparation. If the OD_{600} is higher, cells will get crowded, which makes the analysis difficult. In addition, having more cells in the dish will exhaust the media faster, thus, faster cell death and photobleaching.
17. The time limit of the imaging depends on several factors. First, after a certain time, due to cell division, the imaging field gets crowded, which makes analysis difficult. Second, ConA gets exhausted upon long incubation, and cells start detaching. Third, due to photobleaching, the signal decreases over time. Fourth, prolonged imaging may cause phototoxicity that may kill your cells.

18. During this step and the next ones, use the micropipette tip on the edges of the glass-bottomed well to avoid scratching the center of the glass.
19. To allow efficient attachment of cells on ConA, avoid shaking or disturbing the dish. Keep the cells and the dish at a fixed temperature to avoid stressing cells and to speed up thermal equilibration with the microscope chamber.
20. This step ensures that the incubation chamber is heated to the desired temperature and the microscope stage, as well as the objective lenses, are equilibrated with the incubation chamber. It usually takes more time to equilibrate the entire system than heating the incubation chamber. Because there is usually no direct way to measure the temperature of the lenses and the stage, we recommend starting to heat the whole chamber 2 h in advance of the experiment. If not equilibrated well, the focal plane may change during the experiment, which will cause out-of-focus images.
21. While the microscope incubation chamber is heating, keep the microscope room cold under air conditioning control. This is critical to avoid heating the camera during the course of the experiment, even though most cameras have their own cooling system.
22. Leave the dish open inside the microscope chamber to avoid condensation, which may get the cells out of focus.
23. It is important to incubate the dish at the microscope stage to provide equilibration of the dish, the cells, and the culture media with the environment of the climate chamber.
24. In our experiments, we used 13 stacks with 0.3 μm thickness to cover 3.6 μm total thickness, which is close to the thickness of haploid yeast.
25. You need to decide on light intensity and exposure time for observation of the mitotic spindle. While doing so, you should prioritize the lowest light intensity and exposure time possible to decrease photobleaching and avoid phototoxicity.
26. Select positions with an optimum number of cells. Avoid too crowded areas. Select positions that contain the greatest number of small-budded cells. This will ensure that the cells will reach close to the end of metaphase by the time imaging acquisition starts. Taking many stage positions will allow you to obtain time-lapse movies of many fields of view in the same experiment. However, imaging too many stage positions will also cause faster bleaching of your specimen in a widefield fluorescence microscope. Therefore, reduce the number of stage positions to image if you are experiencing too much

photobleaching. In our lab, we take 4–5 stage positions for *GFP-TUB1* time-lapse experiments.

27. For wild-type haploid budding yeast cells at 30 °C in SC-complete media, anaphase duration is ~18 min. We take 1 h-long time-lapse movies to observe enough cells that go through anaphase during the course of the experiment. If you use mutant cells where anaphase is longer, consider imaging longer.
28. If cells lose focus, and if your microscope acquisition software allows, re-adjust the focus of cells during the experiment without interfering with the time interval between the time-lapse time point.
29. In some situations, the maximum projection works better, so compare both before starting the analysis.
30. Before anaphase onset, for many time points, the spindle length of cells will fluctuate between 1 and 2 μm (Fig. 1). At this stage, the spindle can go back and forth between the mother and daughter cell compartments or rotate in the cell. After cohesin cleavage, the spindle starts elongating rapidly. This can be noticed during the time-lapse as a time point after which the spindle grows in length rapidly. This time point is referred to as the point of metaphase-to-anaphase transition (Fig. 1). After the rapid elongation, the spindle growth rate decreases. Eventually, the spindle reaches a maximum length and stays at this size for some time points. After mitotic exit, the spindle breaks at the spindle midzone, after which the nuclear microtubules shrink (Fig. 1). At this time point, spindle poles usually get closer to each other, and after this point they start moving independently. See Fig. 1 to see how the distance between the spindle poles changes from metaphase through anaphase and after spindle breakdown.
31. Single cells behave differently from each other. To capture the population behavior accurately, 20–100 cells must be analyzed depending on the sample's standard deviation.
32. Fluctuations in the temperature or slight differences in the culture media may cause variations in anaphase duration. Therefore, when comparing anaphase duration between different yeast strains, use the same media and microscope setup and, if possible, do the time-lapse experiments on the same day. If comparing data from experiments performed on different days, include a reference yeast strain (i.e., wild-type cells) on each day and ensure the reference strain's anaphase duration behaves the same on different days.

References

1. Matellán L, Monje-Casas F (2020) Regulation of mitotic exit by cell cycle checkpoints: lessons from Saccharomyces cerevisiae. Genes 11(2): 195. https://doi.org/10.3390/genes11020195
2. Holder J, Poser E, Barr FA (2019) Getting out of mitosis: spatial and temporal control of mitotic exit and cytokinesis by PP1 and PP2A. FEBS Lett 593(20):2908–2924. https://doi.org/10.1002/1873-3468.13595
3. Howell RSM, Klemm C, Thorpe PH, Csikasz-Nagy A (2020) Unifying the mechanism of mitotic exit control in a spatiotemporal logical model. PLoS Biol 18(11):e3000917. https://doi.org/10.1371/journal.pbio.3000917
4. Peters JM (2002) The anaphase-promoting complex: proteolysis in mitosis and beyond. Mol Cell 9(5):931–943. https://doi.org/10.1016/s1097-2765(02)00540-3
5. Yu H (2007) Cdc20: a WD40 activator for a cell cycle degradation machine. Mol Cell 27(1):3–16. https://doi.org/10.1016/j.molcel.2007.06.009
6. Lara-Gonzalez P, Pines J, Desai A (2021) Spindle assembly checkpoint activation and silencing at kinetochores. Semin Cell Dev Biol 117: 86–98. https://doi.org/10.1016/j.semcdb.2021.06.009
7. Baumer M, Braus GH, Irniger S (2000) Two different modes of cyclin clb2 proteolysis during mitosis in Saccharomyces cerevisiae. FEBS Lett 468(2–3):142–148. https://doi.org/10.1016/s0014-5793(00)01208-4
8. Yeong FM, Lim HH, Padmashree CG, Surana U (2000) Exit from mitosis in budding yeast: biphasic inactivation of the Cdc28-Clb2 mitotic kinase and the role of Cdc20. Mol Cell 5(3):501–511. https://doi.org/10.1016/s1097-2765(00)80444-x
9. Visintin R, Prinz S, Amon A (1997) CDC20 and CDH1: a family of substrate-specific activators of APC-dependent proteolysis. Science 278(5337):460–463. https://doi.org/10.1126/science.278.5337.460
10. Visintin R, Craig K, Hwang ES, Prinz S, Tyers M, Amon A (1998) The phosphatase Cdc14 triggers mitotic exit by reversal of Cdk-dependent phosphorylation. Mol Cell 2(6):709–718. https://doi.org/10.1016/s1097-2765(00)80286-5
11. Manzano-Lopez J, Monje-Casas F (2020) The multiple roles of the Cdc14 phosphatase in cell cycle control. Int J Mol Sci 21(3). https://doi.org/10.3390/ijms21030709
12. Stegmeier F, Visintin R, Amon A (2002) Separase, polo kinase, the kinetochore protein Slk19, and Spo12 function in a network that controls Cdc14 localization during early anaphase. Cell 108(2):207–220. https://doi.org/10.1016/s0092-8674(02)00618-9
13. D'Amours D, Stegmeier F, Amon A (2004) Cdc14 and condensin control the dissolution of cohesin-independent chromosome linkages at repeated DNA. Cell 117(4):455–469. https://doi.org/10.1016/s0092-8674(04)00413-1
14. Sullivan M, Higuchi T, Katis VL, Uhlmann F (2004) Cdc14 phosphatase induces rDNA condensation and resolves cohesin-independent cohesion during budding yeast anaphase. Cell 117(4):471–482. https://doi.org/10.1016/s0092-8674(04)00415-5
15. Pereira G, Schiebel E (2003) Separase regulates INCENP-Aurora B anaphase spindle function through Cdc14. Science 302(5653): 2120–2124. https://doi.org/10.1126/science.1091936
16. Higuchi T, Uhlmann F (2005) Stabilization of microtubule dynamics at anaphase onset promotes chromosome segregation. Nature 433(7022):171–176. https://doi.org/10.1038/nature03240
17. Jaspersen SL, Morgan DO (2000) Cdc14 activates cdc15 to promote mitotic exit in budding yeast. Curr Biol 10(10):615–618. https://doi.org/10.1016/s0960-9822(00)00491-7
18. Konig C, Maekawa H, Schiebel E (2010) Mutual regulation of cyclin-dependent kinase and the mitotic exit network. J Cell Biol 188(3):351–368. https://doi.org/10.1083/jcb.200911128
19. Caydasi AK, Khmelinskii A, Duenas-Sanchez R, Kurtulmus B, Knop M, Pereira G (2017) Temporal and compartment-specific signals coordinate mitotic exit with spindle position. Nat Commun 8:14129. https://doi.org/10.1038/ncomms14129
20. Jaspersen SL, Charles JF, Tinker-Kulberg RL, Morgan DO (1998) A late mitotic regulatory network controlling cyclin destruction in Saccharomyces cerevisiae. Mol Biol Cell 9(10): 2803–2817. https://doi.org/10.1091/mbc.9.10.2803
21. Shou W, Seol JH, Shevchenko A, Baskerville C, Moazed D, Chen ZW et al (1999) Exit from mitosis is triggered by Tem1-dependent release of the protein phosphatase Cdc14 from nucleolar RENT complex. Cell 97(2):233–244.

https://doi.org/10.1016/s0092-8674(00)80733-3

22. Caydasi AK, Ibrahim B, Pereira G (2010) Monitoring spindle orientation: spindle position checkpoint in charge. Cell Div 5:28. https://doi.org/10.1186/1747-1028-5-28
23. Stegmeier F, Amon A (2004) Closing mitosis: the functions of the Cdc14 phosphatase and its regulation. Annu Rev Genet 38(1):203–232. https://doi.org/10.1146/annurev.genet.38.072902.093051
24. Huda M, Bektas SN, Bekdas B, Caydasi AK (2023) The signalling lipid PI3,5P(2) is essential for timely mitotic exit. Open Biol 13(9): 230125. https://doi.org/10.1098/rsob.230125
25. McNally FJ (2013) Mechanisms of spindle positioning. J Cell Biol 200(2):131–140. https://doi.org/10.1083/jcb.201210007
26. Caydasi AK, Khmelinskii A, Darieva Z, Kurtulmus B, Knop M, Pereira G (2023) SWR1 chromatin remodeling complex prevents mitotic slippage during spindle position checkpoint arrest. Mol Biol Cell 34(2):ar11. https://doi.org/10.1091/mbc.E20-03-0179
27. Caydasi AK, Kurtulmus B, Orrico MI, Hofmann A, Ibrahim B, Pereira G (2010) Elm1 kinase activates the spindle position checkpoint kinase Kin4. J Cell Biol 190(6): 975–989. https://doi.org/10.1083/jcb.201006151
28. Caydasi AK, Micoogullari Y, Kurtulmus B, Palani S, Pereira G (2014) The 14-3-3 protein Bmh1 functions in the spindle position checkpoint by breaking Bfa1 asymmetry at yeast centrosomes. Mol Biol Cell 25(14):2143–2151. https://doi.org/10.1091/mbc.E14-04-0890
29. Kocakaplan D, Karaburk H, Dilege C, Kirdok I, Bektas SN, Caydasi AK (2021) Protein phosphatase 1 in association with Bud14 inhibits mitotic exit in Saccharomyces cerevisiae. elife 10:10. https://doi.org/10.7554/eLife.72833
30. Markus SM, Omer S, Baranowski K, Lee WL (2015) Improved plasmids for fluorescent protein tagging of microtubules in *Saccharomyces cerevisiae*. Traffic 16(7):773–786. https://doi.org/10.1111/tra.12276
31. Straight AF, Marshall WF, Sedat JW, Murray AW (1997) Mitosis in living budding yeast: anaphase a but no metaphase plate. Science 277(5325):574–578. https://doi.org/10.1126/science.277.5325.574
32. Pereira G, Schiebel E (2001) The role of the yeast spindle pole body and the mammalian centrosome in regulating late mitotic events. Curr Opin Cell Biol 13(6):762–769. https://doi.org/10.1016/s0955-0674(00)00281-7
33. Sherman F (2002) Getting started with yeast. Elsevier, pp 3–41

Chapter 7

Visualizing Cartwheel Disassembly Process During Mitosis in Fixed and Live Cells by Fluorescence Microscope

Fan Huang and Chuanmao Zhang

Abstract

Centrosome is an evolutionarily conserved organelle that comprises two barrel-shaped centrioles surrounded by pericentriolar material (PCM). It functions as the major microtubule-organizing center (MTOC) to regulate cell polarity, motility, intracellular material transport during interphase, and bipolar spindle assembly during mitosis. Cartwheel assembly is considered the first step in the initiation of procentriole biogenesis at early S phase. In human cells, cartwheel is a transient scaffolding structure that is disassembled during procentriole maturation at mitotic exit. This chapter describes methods for synchronizing mammalian cells into mitosis by small molecule inhibitor treatment and visualizing cartwheel disassembly process during mitosis in fixed and live cells by fluorescence microscope.

Key words Centrosome, Centriole, MTOC, Cartwheel disassembly, Mitotic exit, Mitosis, Fluorescence microscope

1 Introduction

Centrosome comprises two barrel-shaped centrioles surrounded by pericentriolar matrix (PCM), and each centriole is characterized by ninefold radial symmetric arrangement of microtubules [1]. A structure called cartwheel serves as a scaffold which characterized by a central hub from which nine spokes project into outer microtubules to support centriole arrangement during its biogenesis. Sas-6 protein, in particular, constitutes the primary backbone of the cartwheel, which forms homodimers via the coiled-coil domains. Nine adjacent homodimers can self-oligomerize via the N-terminal head domain to form a ring resembling the central hub, with the C-terminal tails oriented outward as spokes [2–7]. Other centriolar proteins like STIL, Cep135 as well as CPAP also functionally cooperates with Sas-6 to facilitate cartwheel assembly, outer microtubules binding and maintain procentriole structure during procentriole biogenesis [1].

Victor M. Bolanos-Garcia (ed.), *Mitotic Exit: Methods and Protocols*, Methods in Molecular Biology, vol. 2874, https://doi.org/10.1007/978-1-0716-4236-8_7,

In human cells, the Sas-6-assembled cartwheel is a transient scaffolding structure that is disassembled during procentriole maturation at mitotic exit [8]. During early S phase, each of the disengaged centrioles acts as a mother centriole to recruit Sas-6 for initiating cartwheel assembly at its lateral proximal end, followed by microtubule nucleation around the cartwheel for procentriole biogenesis [9–11]. Then Sas-6 is continuously recruited and immobilized to the procentriole from S phase to late G2 phase, leading to an increase in the size of the cartwheel structure. Upon cell entering into mitosis, Sas-6-assembled cartwheel is dissembled and disappeared from centrioles gradually [12]. Based on our recent research findings, Sas-6-assembled cartwheel has been shown to serve as a linker to maintain engagement of paired centrioles after centriole biogenesis, and cartwheel disassembly regulated by CDK1-cyclin B kinase during mitosis allows centriole disengagement and licensing [13].

This chapter describes detailed methods for synchronizing mammalian cells into mitosis by small molecule inhibitor treatment and visualizing cartwheel disassembly process during mitosis in fixed and live cells by fluorescence microscope. To generate mammalian cells freshly exited into mitosis, cells should be firstly synchronized at G1/S transition using double-thymidine blockade method, then releases cells into fresh culture medium for 9.5 h to mitosis. The corresponding protocol is described in Subheading 3.1. Through staining DNA with DAPI to recognize cell cycle stages and using antibodies against Centrin1 and Sas-6 to indicate centriole and cartwheel separately, it is possible to observe cartwheel disassembly process during mitosis by fluorescence microscope in fixed cells. The corresponding protocol is described in Subheading 3.2. Live-cell imaging of cells stably expressing GFP-Sas-6 and transiently expressing RFP-H2B visualizes process of cartwheel disassembly during mitosis progression within live cells. The corresponding protocol is described in Subheading 3.3.

2 Materials

2.1 Synchronizing Cultured Mammalian Cells into Mitosis

1. Cell lines of interest (e.g., wild-type HeLa cell, wild-type U2OS cells, HeLa cells stably expressing GFP-Sas-6).
2. Culture medium: Dulbecco's Modified Eagle's Medium (DMEM) supplemented with 10% fetal bovine serum (FBS), 1% Penicillin-Streptomycin.
3. 0.05% trypsin–EDTA.
4. Phosphate-buffer saline (PBS) solution: 135 mM NaCl, 4.7 mM KCl, 10 mM Na_2HPO_4, 2 mM NaH_2PO_4, PH 7.4.
5. Dimethyl sulfoxide (DMSO).

6. Thymidine: a DNA synthesis inhibitor for synchronization of cells in G1/S transition, 250 mM stock solution in DMSO, stored at −20 °C.
7. RO3306: a selective CDK1 inhibitor, 9 mM stock solution in DMSO, stored at −20 °C.

2.2 Observation of Cartwheel Disassembly in Fixed Cells by Fluorescence Microscope

1. Cell lines of interest (e.g., wild-type HeLa cell, wild-type U2OS cells).
2. 3.5 cm cell culture dishes, glass coverslips, and microscope slides.
3. Fixative buffer: methanol stored at −20 °C.
4. Antibody dilution buffer: 3% BSA dissolved in PBS solution, stored at −20 °C.
5. Antibodies: mouse anti-Sas-6 (Santa Cruz, sc-81431), rabbit anti-Centrin1 (Proteintech, 12794-1-AP), and appropriate secondary antibodies.
6. Mounting medium: 1 μg/mL DAPI and 2.5% DABCO dissolved in Mowiol, stored at −20 °C.
7. Nail polish.
8. Fluorescence microscope (e.g., Delta Vision Elite microscope, High-speed spinning-disk confocal microscope).

2.3 Observation of Cartwheel Disassembly in Live Cells by Fluorescence Microscope

1. Cell lines of interest (e.g., GFP-Sas-6-expressing HeLa cell lines).

   The Stable GFP-Sas-6-expressing HeLa cell lines were generated via lentivirus infection (*see* **Note 1**). Briefly, the lentivirus transfer plasmids pCDH-CMV-MCS-EF1-Puro-GFP-Sas-6 were introduced into HEK293T cells together with packing plasmids psPAX2 and envelop plasmids pMD2G. The generated recombinant lentivirus was then used to infect HeLa cells, and the cell population expressing GFP-Sas-6 was selected using puromycin.
2. 3.5 cm glass-bottom cell culture dishes.
3. Plasmids encoding RFP-H2B.
4. Lipofectamine 2000 reagent.
5. Fluorescence microscope (e.g., Delta Vision Elite microscope, High-speed spinning-disk confocal microscope).

3 Methods

In human cells, Sas-6-assembled cartwheel is a transient scaffolding structure that exists on centriole from early S phase to early mitosis. Upon entering mitosis, the cartwheel disassembles and disappears gradually from centriole. Thus, the discrimination of cell cycle stage and application of suitable centriole and cartwheel marker are the core factors to visualize the cartwheel disassembly process during mitosis.

Mitosis comprises four distinct phases: prophase, metaphase, anaphase, and telophase. Each distinct phase can be easily judged by visual inspection of chromosome morphology via DAPI staining in fixed cells or transient expression of RFP-H2B in live cells. The main occurrences in prophase are the condensation of the chromatin and the disappearance of the nucleolus. In metaphase, these compact chromosomes align in the equatorial plate, an imaginary line that is equidistant from the two centrosome poles. The duplicated chromosomes later are split and pulled toward the opposite poles of the cells as entering anaphase. During telophase, the separated chromosomes reach the cell poles, the chromosomes begin to decondense back into the expanded chromatin and the nucleolus reappear.

Normally have identified the cell cycle stage, the next step is to visualize the centriole and cartwheel structure. In fixed cell, staining pretreated cells with antibodies against Centrin1 and Sas-6 to indicate centriole and cartwheel separately. Centrosome with assembled cartwheel is determined by two centrioles (two Centrin1 dots) and one cartwheel (one Sas-6 dots), the Sas-6 signal weakens and disappears along with cartwheel disassembly as cell entering mitosis (Fig. 1). In live cells, the cartwheel disassembly process can be directly observed via real-time tracing the GFP-Sas-6 signal during mitosis (Fig. 2).

3.1 Synchronizing Cultured Mammalian Cells into Mitosis

1. Plate cells in a culture dish. For fixed cell imaging, plate the HeLa cells with 70% confluence on sets of coverslips placed in 3.5 cm cell culture dishes (*see* **Notes 2** and **3**), then incubate the cells in a cell culture incubator with 5% CO_2 for 12 h to let the cells settle and adhere. For live-cell imaging, plate the stable GFP-Sas-6-expressing HeLa cell lines with 50% confluence in glass-bottom cell culture dishes, then transfect cells with RFP-H2B plasmids for 24 h after the cells have settled and adhered.
2. Treat cells with 2.5 mM Thymidine for 17–21 h to arrest cell in G1/S and S phase (*see* **Note 4**).
3. Replace the medium with fresh culture medium for 10 h to release cells into subsequent cell cycle stages (*see* **Note 5**).
4. Retreat cells with 2.5 mM Thymidine for 17–21 h to arrest cell at G1/S transition (*see* **Note 4**).
5. Replace the medium with fresh culture medium for 9.5 h to release cells into mitosis.

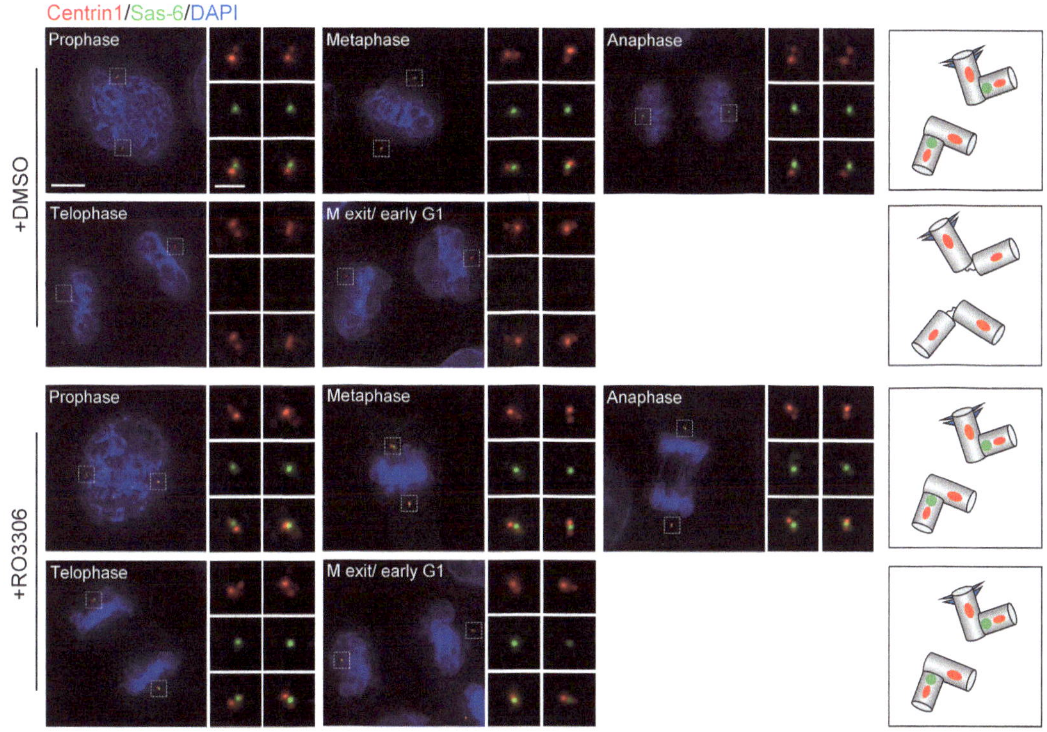

Fig. 1 Observation of cartwheel disassembly in fixed cells by fluorescence microscope. Sas-6-assembled cartwheel disassembled from centriole during late mitosis and inhibition of CDK1-cyclin B activity prevents cartwheel disassembly during late mitosis. HeLa cells were arrested at G1/S transition by double-thymidine block method, released for 9.5 h to mitosis, and treated with DMSO or RO3306 for 15 min. HeLa cells were fixed with cold methanol and analyzed by immunofluorescence using antibodies against Sas-6 and Centrin1. DNA was stained with DAPI. Scale bars, 5 μm (large images) or 1 μm (inset images)

3.2 Observation of Cartwheel Disassembly in Fixed Cells by Fluorescence Microscope (Fig. 1)

1. To study how CDK1 kinase activity influences the cartwheel disassembly, cells are treated with DMSO or selective CDK1 inhibitor RO3306 (9 μM working concentration) for 15 min before fixation (*see* **Note 6**).
2. Remove the culture medium and wash cells with PBS at room temperature (*see* **Note 6**).
3. Fix the cells with −20 °C methanol on ice for 5 min.
4. Remove the remained fixation buffer and wash the cells three times with ice-cold PBS.
5. Incubate the fixed cells with the primary antibody (anti-Sas-6, anti-Centrin1 antibodies) diluted in 3% BSA in a humidified chamber overnight at 4 °C (*see* **Note 7**).
6. Decant the primary antibody solution and wash the cells three times with PBS, 5 min each wash.
7. Incubate the fixed cells with secondary antibody (Alexa Fluor 488-conjugated donkey anti-mouse IgG, Alexa Fluor

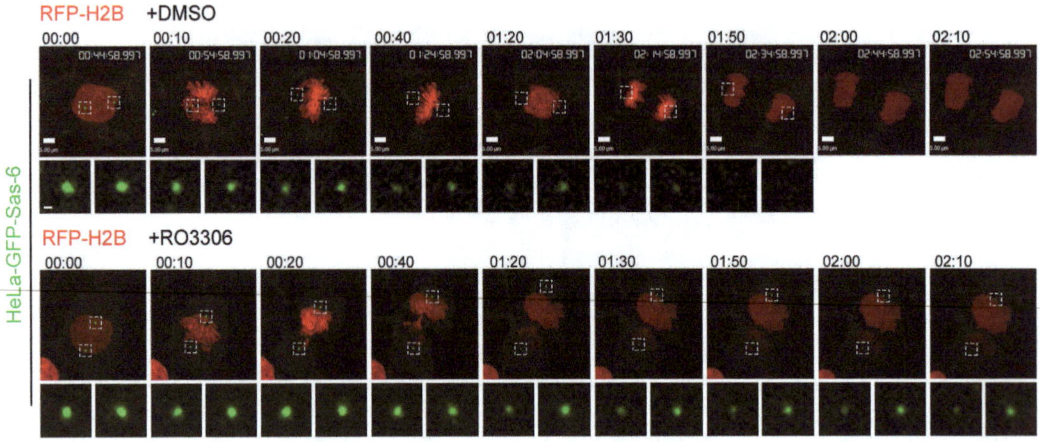

Fig. 2 Observation of cartwheel disassembly in live cells by fluorescence microscope. GFP-Sas-6 signal weaken and disappear from centriole during mitosis, and inhibition of CDK1-cyclin B activity prevents this process. HeLa cells stably expressing GFP-Sas-6 were transfected with RFP-H2B, then treated with DMSO or RO3306. The movies were filmed for 3 h from late G2 phase to mitotic exit. Images were captured every 5 min, and 30 slices at an 18-μm Z-section thickness were obtained. Scale bar, 5 μm (large images) or 1 μm (inset images)

594-conjugated donkey anti-rabbit IgG) diluted in 3% BSA for 1 h at room temperature in the dark.

8. Decant the secondary antibody solution and wash the cells three times with PBS, 5 min each wash.
9. Preheat and melt the mounting medium on thermal shaker (*see* **Note 8**).
10. Mount glass coverslips with a drop of melted mounting medium on microscope slides (*see* **Note 9**).
11. Let the microscope slides rest in dark for 1 h at room temperature to let the mounting medium solidify, then seal glass coverslips with nail polish to prevent drying and movement under microscope.
12. Store the glass slides in dark at 4 °C for several days if not observe immediately (*see* **Note 10**).
13. Using the cell imaging system on fluorescence microscope to visualize chromosome (indicated by DAPI staining), centriole (indicated by Centrin1 staining) and cartwheel (indicated by Sas-6 staining) with an excitation wavelength of 350 nm, 594 nm, and 488 nm respectively.
14. Search the cells first at low magnification in DAPI channel then focus on individual mitotic cells at high magnification.
15. To observe the cartwheel structure at different mitotic stages, centrioles and cartwheels were visualized via staining Centrin1 and Sas-6 antibodies. Normally Centrin1 localizes to the

centriole distal lumen and Sas-6 localizes on the proximal of procentrioles (Fig. 1).

16. Images are captured at a 4-μm Z-section thickness in 20 slices for indicated mitotic cells. The data of captured image are processed for maximum intensity projection using the Volocity software.

3.3 Observation of Cartwheel Disassembly in Live Cells by Fluorescence Microscope (Fig. 2)

1. Using the live-cell imaging system on fluorescence microscope to visualize chromosome (indicated by RFP-H2B signal) and cartwheel (indicated by GFP-Sas-6 signal) with an excitation wavelength of 594 nm and 488 nm, respectively.
2. Search the cells first at low magnification in TRITC channel then focus on individual late G2 phase cells at high magnification.
3. Select ten suitable cells for imaging. The most suitable cells are that at late G2 phase and are going to prophase, these cells exhibit enlarged nuclear (indicated by RFP-H2B signal) and separated centrosomes (indicated by GFP-Sas-6 signal) which locate at opposite edges of nuclear.
4. Images were captured at an 18-μm Z-section thickness in 30 slices every 5 min.
5. To study how CDK1 kinase activity influences the cartwheel disassembly, cells newly entered mitosis should be immediately treated with DMSO or selective CDK1 inhibitor RO3306 (working concentration 9 μM) in time-lapse intervals during image capture (*see* **Note 11**). To ensure that the tracked cells have completed mitosis and entered next G1 phase, the movies of the DMSO/RO3306 treated cells are recorded for 3 h by fluorescent microscope system.
6. The data of captured image are processed using the Fusion software.

4 Notes

1. To generate stable GFP-Sas-6-expressing HeLa cell lines via lentivirus infection. This method can introduce low GFP-Sas-6 expression in cells to avoid inducing centriole over-duplication.
2. U2OS and HeLa are the cell lines that are commonly used to visualize centrosome. The U2OS and HeLa cells are sensitive to cell cycle synchronizing drug. Moreover, the centrosomes generated in U2OS and HeLa cells are usually large and easy to visualize the partial organization of paired centrioles.

3. The cell density should keep within 70–80% before thymidine treatment, thus to maintain a good cell status while at the same time provide the cells with enough space for division.
4. Thymidine is a DNA synthesis inhibitor that arrests cells at the G1/S transition prior to DNA replication. Double-thymidine block can improve the synchronization efficacy. The first thymidine block arrests cells in the G1/S transition and different stages of S phases, then the second thymidine block synchronizes all the cells to the G1/S transition.
5. When releasing thymidine-treated cells, the cells should be washing by PBS twice carefully to remove the remaining thymidine, then add the fresh culture medium to ensure cells enter subsequent cell cycle stages.
6. When adding compounds into or washing the mitotic-arrested cells, gently aspirate and pipette the medium or PBS to avoid loss of mitotic cells for their weak adherence.
7. The incubation time for fixed cells with the primary antibody can extent to several days at 4 °C, if it cannot proceed to the next step in time. Remember to keep humidified chamber hydrated and avoid drying out of glass coverslips.
8. The mounting medium should be stored at −20 °C for long times, and thaw immediately before use.
9. When mounting glass coverslips, remove the excess mounting medium and carefully put coverslips on it to avoid introducing air bubbles between the slide and coverslip.
10. Once made, the glass slides can be stored at 4 °C for few months if they are kept moist, although typically observed immediately.
11. RO3306 should be immediately added into medium upon cells entering mitosis. However, the treated cells cannot enter mitosis if RO3306 is added too early, or cartwheel disassembly process has begun if RO3306 is added too late. Moreover, in order to capture the GFP-Sas-6 signal in time after DMSO/RO3306 treatment, cells of interest should be traced firstly by image capture program from late G2 phase, then gently add DMSO/RO3306 into newly mitotic cells in time-lapse intervals during image capture. Be careful to avoid movement of glass-bottom cell culture dishes during dosing.

References

1. Nigg EA, Holland AJ (2018) Once and only once: mechanisms of centriole duplication and their deregulation in disease. Nat Rev Mol Cell Biol 19(5):297–312. https://doi.org/10.1038/nrm.2017.127
2. Guichard P et al (2012) Cartwheel architecture of Trichonympha basal body. Science 337(6094):553. https://doi.org/10.1126/science.1222789

3. Hilbert M et al (2016) SAS-6 engineering reveals interdependence between cartwheel and microtubules in determining centriole architecture. Nat Cell Biol 18(4):393–403. https://doi.org/10.1038/ncb3329
4. Kitagawa D et al (2011) Structural basis of the 9-fold symmetry of centrioles. Cell 144(3): 364–375. https://doi.org/10.1016/j.cell.2011.01.008
5. Nakazawa Y, Hiraki M, Kamiya R, Hirono M (2007) SAS-6 is a cartwheel protein that establishes the 9-fold symmetry of the centriole. Curr Biol 17(24):2169–2174. https://doi.org/10.1016/j.cub.2007.11.046
6. van Breugel M et al (2011) Structures of SAS-6 suggest its organization in centrioles. Science 331(6021):1196–1199. https://doi.org/10.1126/science.1199325
7. van Breugel M et al (2014) Structure of the SAS-6 cartwheel hub from Leishmania major. elife 3:e01812. https://doi.org/10.7554/eLife.01812
8. Strnad P et al (2007) Regulated HsSAS-6 levels ensure formation of a single procentriole per centriole during the centrosome duplication cycle. Dev Cell 13(2):203–213. https://doi.org/10.1016/j.devcel.2007.07.004
9. Carvalho-Santos Z et al (2010) Stepwise evolution of the centriole-assembly pathway. J Cell Sci 123(Pt 9):1414–1426. https://doi.org/10.1242/jcs.064931
10. Conduit PT, Wainman A, Raff JW (2015) Centrosome function and assembly in animal cells. Nat Rev Mol Cell Biol 16(10):611–624. https://doi.org/10.1038/nrm4062
11. Fu J, Hagan IM, Glover DM (2015) The centrosome and its duplication cycle. Cold Spring Harb Perspect Biol 7(2):a015800. https://doi.org/10.1101/cshperspect.a015800
12. Keller D et al (2014) Mechanisms of HsSAS-6 assembly promoting centriole formation in human cells. J Cell Biol 204(5):697–712. https://doi.org/10.1083/jcb.201307049
13. Huang F et al (2022) Cartwheel disassembly regulated by CDK1-cyclin B kinase allows human centriole disengagement and licensing. J Biol Chem 298(12):102658. https://doi.org/10.1016/j.jbc.2022.102658

Chapter 8

FRET-FLIM for the Study of Protein-Protein Interactions Underpinning Mitosis Checkpoints

Charlotte Pain

Abstract

Cell division is a key cellular process that ensures the continuation of life on Earth. In order to protect the genetic integrity of organisms, cell division must happen accurately, ensuring each daughter cell receives a complete copy of the original genome. The accuracy of this process is, in part, preserved by various cell cycle checkpoints. These checkpoints rely on the physical interactions of their components to ensure proper function. The spindle assembly checkpoint (SAC), for example, produces an inhibitory complex of BUBR1-BUB3 and MAD2 bound to CDC20. Many of these cell cycle checkpoint components have been identified in plants, but it has not yet been established whether plants have a mitotic checkpoint architecture that is similar to mammalian cells. To understand the function of plant cell cycle homologues, it is imperative to characterize their interactions in vivo. FRET-FLIM (Förster resonance energy transfer-fluorescence lifetime imaging microscopy), is a rapidly expanding technique that can be used to rapidly and simply characterize protein-protein interactions.

Key words Spindle assembly checkpoint, FRET-FLIM, Fluorescence imaging

1 Introduction

The spindle assembly checkpoint (SAC) is an essential cell cycle checkpoint, controlling cell cycle progression to anaphase until all chromosomes are correctly attached to the spindle prior to movement of chromatids to each end of the cell. The SAC of *Arabidopsis thaliana* (thale cress) is composed of multiple components, including BUB1, BUB3.1, BUB3.2, BUB3.3 (budding uninhibited by benzimidazole), MAD1, MAD2, MAD3.1, MAD3.2 (mitotic-arrest deficient), and MPS1 in plants (monopolar spindle) [1]. The SAC senses the status of the kinetochores and while the kinetochores remain unattached, promotes a "wait anaphase" signal though the anaphase promoting complex/cyclosome (APC/C). In order to signal correctly, essential stable protein complexes must be formed. The SAC "wait anaphase" signal is transmitted through an inhibitory complex produced by the SAC prior to

Victor M. Bolanos-Garcia (ed.), *Mitotic Exit: Methods and Protocols*, Methods in Molecular Biology, vol. 2874,
https://doi.org/10.1007/978-1-0716-4236-8_8,

complete kinetochore attachment. This complex is the mitotic checkpoint complex (MCC) and is composed of MAD2 and BUBR1–BUB3 bound to CDC20 (cell cycle division 20) [2, 3]. For this complex to function properly, it is essential that MAD2 and CDC20 (the APC/C activator) interact directly [4–9]. If this interaction is prevented, for example when expressing CDC20 mutants that have impaired MAD2 binding, the SAC is no longer functional and post-anaphase processes were shown to occur in yeast even when the spindle was not fully assembled [6, 8]. The interaction of BUBR1 and BUB1 is also required for their localization to the kinetochore [10–12]. The function of the SAC as a cell cycle checkpoint is therefore dependent on the physical interactions of its component parts and other signaling partners. To establish the architecture of the plant SAC and its functional roles, it is essential to understand how the plant SAC components interact with each other.

FRET-FLIM (Försters resonance energy transfer—fluorescence lifetime imaging) is a well-established tool for characterizing protein-protein interactions in vivo [13–16]. The technique relies on the underlying principle that fluorescent molecules with particular spectral characteristics (FRET pairs), will interact with one another when they are in close enough proximity (typically less than 10 nm [17]). By detecting the hallmarks of this interaction, it is then possible to infer that proteins in such close proximity are likely to directly interact with one another. When a fluorescent molecule (the "donor") absorbs energy, for example, when it is targeted by a laser of the correct wavelength, electrons in the molecule will become excited. In the absence of another nearby fluorescent molecule, the electrons will remain in this unstable excited state for a period of time until they return to their unexcited state, releasing this energy as fluorescence [17–20]. The period of time during which the electrons remain in a higher energy state is known as the fluorescence lifetime. When another fluorescent molecule with the required spectral characteristics is nearby however, this energy is more quickly transferred to the neighboring molecule ("the acceptor") in a process known as Försters resonant energy transfer (FRET) [17, 21]. The acceptor can then release this energy itself. As this transfer happens relatively quickly, FRET results in a shorter fluorescence lifetime [22]. FRET alone can be used to study protein-protein interactions, however with improved photon counting abilities it is possible to perform FRET-FLIM. FRET-FLIM is considered superior to FRET alone as the fluorescence lifetime of any given fluorophore is an intrinsic property. Therefore, it is not impacted by factors that can affect the fluorescence intensity of fluorophores, for example, the concentration of a fluorophore [20, 23]. In addition, FLIM produces data that can be resolved spatially, building a more complex picture of the location of interactions [24].

When designing a FRET-FLIM experiment, it is important to consider the following factors: the localization of the fluorescent protein, the position of the fluorescent tag within the protein and the spectral properties of the selected FRET pair. First and foremost, it is important that the addition of fluorescent tags to proteins does not cause protein mislocalization as this may spatially separate the proteins being tested from one another. Some fluorophores are known to be more likely to cause protein mislocalization, so all fluorophores should be tested prior to setting up an experiment [25]. Second, FRET-FLIM can only occur across a short distance, typically only 3–6 nm [26]. Therefore, it is essential that the correct termini of the protein have been tagged in order for FRET to be possible. This requirement for fluorophores to be in close proximity can also be exploited to establish membrane topology of interacting proteins. For example, by adding a fluorescent tag to each termini of particulate methane monooxygenase (pMMO) enzyme complex subunits, it has been possible to establish the membrane topology of the pMMO complex [27]. Finally, it is important to select donor and acceptor fluorophores that can function as FRET pairs on the basis of their spectral characteristics. Key features to consider when selecting a FRET pair are as follows:

- Donor fluorophores must have a low susceptibility to photobleaching.
- The emission spectrum of the donor fluorophore must significantly overlap with the absorption spectrum of the acceptor fluorophore. An overlap of at least 30% is recommended [28].
- Fluorophores must have a sufficient quantum yield.

In addition, different fluorophore combinations will be able to interact across different distances. The distance between two fluorophores across which FRET can be detected is the Förster radius. Specifically, Förster radius is defined as the distance between the donor and acceptor fluorophore for 50% FRET efficiency [19]. This distance can be calculated using the characteristics of the fluorophores chosen for the FRET pairs. These details can be easily extracted from FPBase (an online repository for fluorescent protein details), which also hosts a "FRET Calculator" that can be used to calculate various FRET characteristics, including Förster radii and the spectral overlap of donor emission and acceptor absorbance [19, 29] (Fig. 1).

There are many FRET pairs that have been successfully tested and applied to plant cell research, each in different plant and cell types and at different locations within the cell. The traditional pairing is eGFP and mRFP as these fluorophores are well tested and widely applied across plant cell research. However, as more fluorophores have been developed, additional FRET pairs are being explored (Table 1).

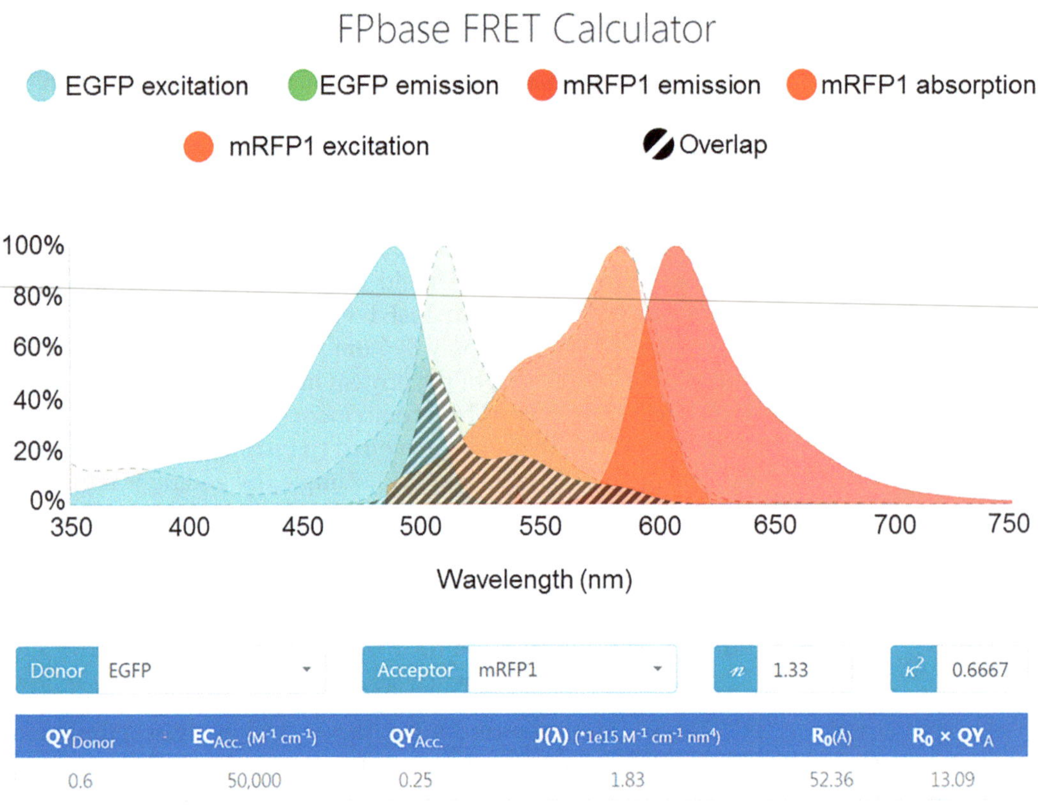

Fig. 1 Example output of the FPbase FRET Calculator, showing the FRET pair calculations for eGFP and mRFP1. Example output from FPBase FRET calculator, which can be used to establish whether two fluorophores are likely to make a suitable FRET pair [19, 29]. The graph shows the excitation, absorption, emission, and overlap of the entered fluorophores in the FRET pair. Key details, including the Förster radius, are included below the graph. Some text has been altered to increase text size and readability

During the testing of more recently developed, theoretically possible fluorophore combinations, several FRET pairs have been identified as being inappropriate for FRET-FLIM experiments. These fluorophores include T-Sapphire and CytERM which are not recommended as a primary choice for FRET-FLIM (Table 2). However, it should be noted that these concerns may be linked to specific experimental setup.

Here I will describe techniques for the analysis of protein-protein interactions by FRET-FLIM in the nuclei of transiently transformed tobacco leaf epidermal cells.

2 Materials

2.1 Performing FRET-FLIM on Tobacco Epidermal Cells

1. Transformed tobacco leaves, expressing appropriate FRET pairs.
2. Razor blades.

Table 1
Fluorophore pairs that are suitable for FRET-FLIM imaging in plant cells

FRET pair					
Donor	**Acceptor**	**R_0(Ä)**	**Localization**	**Species**	**Reference**
eGFP	mRFP	52.36	ER	*Nicotiana tabacum* leaves (transient)	[13]
SCFP3A	SYFP2	N/A	Nuclear	Arabidopsis roots	[30]
mTurquoise	SYFP2	57.31	Nuclear	Arabidopsis roots	[30]
SYFP2	mCherry	58.73	Nuclear	Arabidopsis roots	[30]
SYFP2	mStrawberry	62.41	Nuclear	Arabidopsis roots	[30]
SYFP2	mRFP	55.41	Nuclear	Arabidopsis roots	[30]
mTurquoise2[a]	mVenus[a]	58.5	Nuclear	*Nicotiana benthamiana* leaves (transient)	[31]
mEGFP[a]	mCherry[a]	52.88	Nuclear	*Nicotiana benthamiana* leaves (transient)	[31]
mVenus[a]	tagRFP[a]	61.34	Nuclear	*Nicotiana benthamiana* leaves (transient)	[31]
Venus	mKate2	54.55	PM/ER	*Nicotiana benthamiana* leaves (transient)	[32]
Venus	mRuby3	62.77	PM/ER	*Nicotiana benthamiana* leaves (transient)	[32]
mTurquoise2	mNeonGreen	61.55	PM/ER	*Nicotiana benthamiana* leaves (transient)	[32]
mTurquoise2	Venus	57.62	PM/ER	*Nicotiana benthamiana* leaves (transient)	[32]

Known FRET pairs that have been tested in different plant tissues. Examples are shown with the name of the fluorophore of the donor and acceptor, the calculated Försters radius (R_0), the localization of the probes when tested and the species/cell type the FRET pair has been tested in

[a]denotes 2in1 vectors that can be used to express two protein/fluorophore fusions from a single vector

3. Microscope slides and coverslips.
4. H_2O.
5. Tape (e.g., micropore tape for securing coverslips).
6. Two-channel confocal with FLIM, for example, Nikon TE2000-U inverted microscope with modified Nikon EC2 confocal scanning system. This should be supplied with the following: Mode-locked titanium sapphire laser (e.g., Coherent Lasers, Mira), Solid-state, continuous-wave, 532-nm laser (e.g., Coherent Lasers, Verdi V18), High–numerical aperture objective (e.g., Nikon VC series). BG39 filter (e.g., Comar), 633-nm interference filter and TCSPC card (e.g., Becker and Hickl SPC 830 or SPC150 or equivalent).

Table 2
FRET pairs that have been reported as not appropriate for FRET-FLIM

Donor	Acceptor	Reported issues	Reference
T-sapphire	mOrange	Proteins may mislocalize and there are challenges with detecting T-sapphire signal	[32]
Clover	mCerulean3	Excessive photobleaching	[32]
Clover	mRuby2	Failure to detect FRET	[32]
SYFP2	–	Rapid nonlinear bleaching	[25]
CytERM	–	Protein may mislocalize	[25]
CytERM	mRuby2	Spectral overlap between the two fluorophores	[25]

Several FRET pairs have been tested and have been reported as not appropriate for FRET-FLIM are listed, along with a brief reason for the recommendation

2.2 Analysis of FRET-FLIM Data

1. FLIM image analysis software (e.g., Becker and Hickl SPCImage NG Data Analysis software, v5.1, or the latest version).
2. A computer system with the capability to run the FLIM image analysis software. Minimum requirements for optimal data analysis include a graphic adapter from NVIDIA with CUDA standard, compatible graphics card (e.g., GTX series above 1050, the Quadro series, and the RTX series, *see* **Note 1**).

3 Methods

3.1 Performing FRET-FLIM on Tobacco Leaf Epidermal Cells

1. Cut a small square of transformed tobacco leaf (*see* **Note 2**).
2. Mount the tobacco leaf piece on a microscope slide using clean water, with the abaxial side of the leaf facing the coverslip.
3. Secure the coverslip in place using appropriate tape and place the slide on the microscope (*see* **Note 3**).
4. Using the confocal system, ensure that the FRET pair fluorescence levels are both detectable and sufficient in the area of interest (Fig. 2). Also ensure that the sample is correctly in focus before acquiring FRET-FLIM data.
5. Set the FRET-FLIM system laser to a wavelength of 920 nm, with 200-fs pulses at 76 MHz, using the titanium sapphire laser, pumped by a solid-state continuous wave 532-nm laser.
6. Focus the laser beams on a diffraction limited spot using an objective with a high numerical aperture (NA 1.2 is recommended in this case, e.g., the Nikon x60 VC NA 1.2).

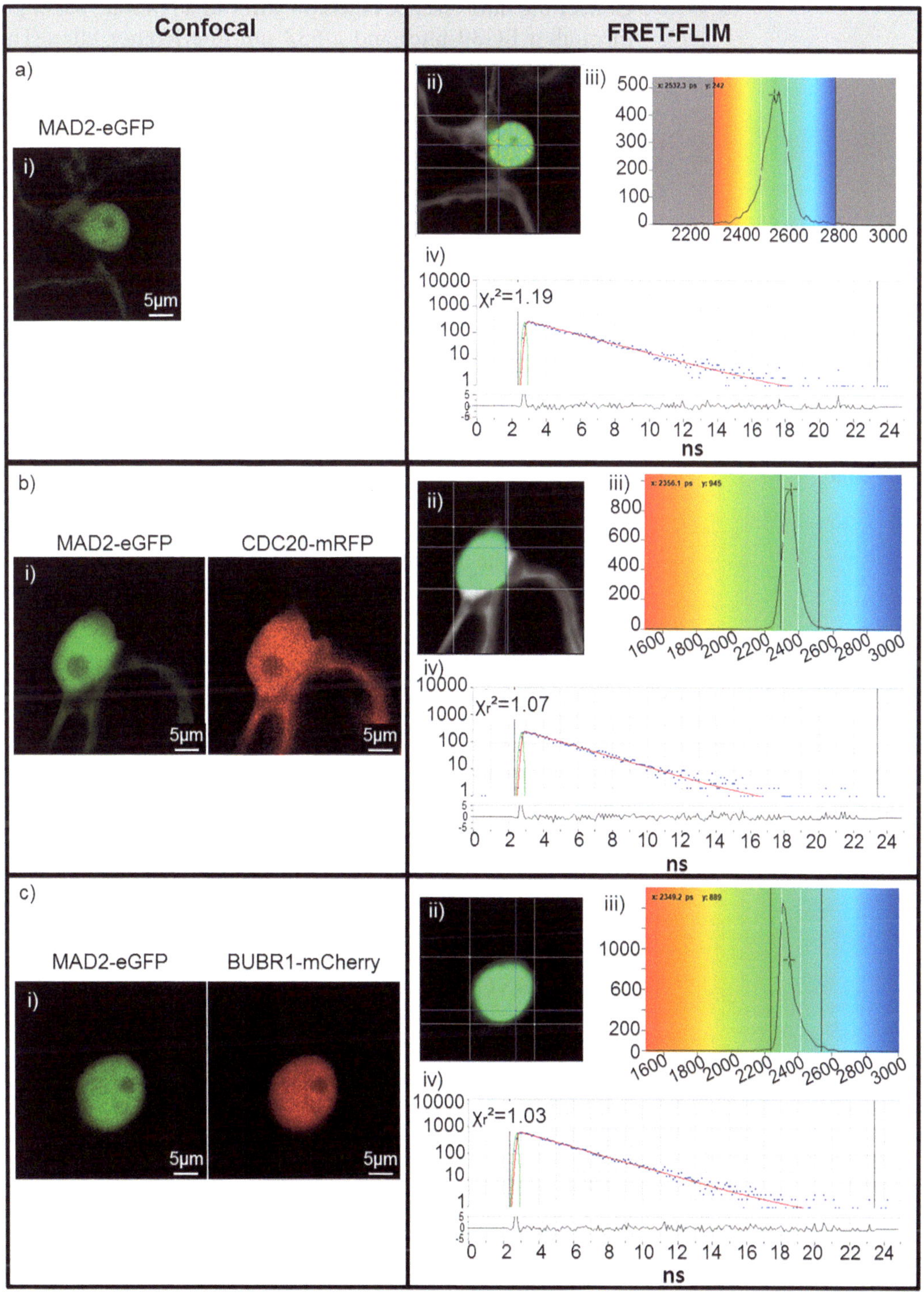

Fig. 2 FRET-FLIM analysis of MAD2-eGFP interactions using SPCImage NG. FRET-FLIM analysis of protein-protein interactions, comparing the fluorescence lifetime of (**a**) MAD2-eGFP alone, (**b**) MAD2-eGFP and CDC20-mRFP and (**c**) MAD2-eGFP and BUBR1-mCherry. (i) Confocal images are shown for each analyzed nuclei. (ii) A psuedocolored fluorescence lifetime map is shown for each nuclei (iii), with only the colored area within the light gray box chosen for analysis. The gray sections of the image are ignored as they do not contain

7. Collect the fluorescence emission without a pinhole, passing it through a BG39 filter and a 633 nm interference filter. This BG39 filter will block the near infrared laser light, while the 633 nm interference filter will block unwanted autofluorescence produced by chlorophyll (*see* **Note 4**).
8. Using the TCSPC PC module SPC830, generate raw, time-correlated single photon data. This data is then used to create FLIM images that can be analyzed using specialist software.
9. Prior to collecting a large dataset, to test the system it is recommended that known positive and negative controls are used to ensure that the system is working well (*see* **Note 5**). Once a control interaction has been tested and confirmed, collect data from different conditions.

3.2 Analysis of FRET-FLIM Data

The SPCImage software fits a model to the fluorescence lifetime imaging and minimizes the X_r^2 between the model and the data. The default and simplest model is a single exponential decay; however, users can select more complex models depending on the requirements of the individual experiment. This can then be used to calculate the fluorescence lifetime of each pixel in the image, which can be interpreted to establish whether a protein-protein interaction has been identified.

1. Load images into the SPCImage NG analysis software.
2. Identify a region of interest (*see* **Note 6**).
3. Generate the lifetime values of each pixel using the software. It will automatically generate a fluorescence lifetime map and will display a distribution curve.
4. Fit the decay data to a single exponential parameter as $f(t)=\alpha\varepsilon^{-T/T}$. Only pixels where the chi-squared value is between 0.9 and 1.4 are considered as having a good fit. Therefore, only pixels with this value should be taken forward for analysis. The SPCImage NG image software allows for filtering on the basis of chi-squared value (*see* **Note 7**). Using the fluorescent lifetime curve created by this analysis (Fig. 2iii), record the median, minimum and maximum lifetime values from each acquisition.
5. Compare the median fluorescent lifetime of each acquisition to identify protein-protein interactions (Fig. 3). Where there is a reduction in FLIM, an interaction between fluorophores has been detected (*see* **Note 8**).

Fig. 2 (continued) a region of interest and may affect the results if analyzed. The distribution of lifetimes across the selected region of interest are also displayed. (iv) A representative decay curve, with an optimum exponential fit for a single point in the image (shown by the dark blue crosshairs). Images were collected from tobacco leaf epidermal cells transiently expressing the protein combinations

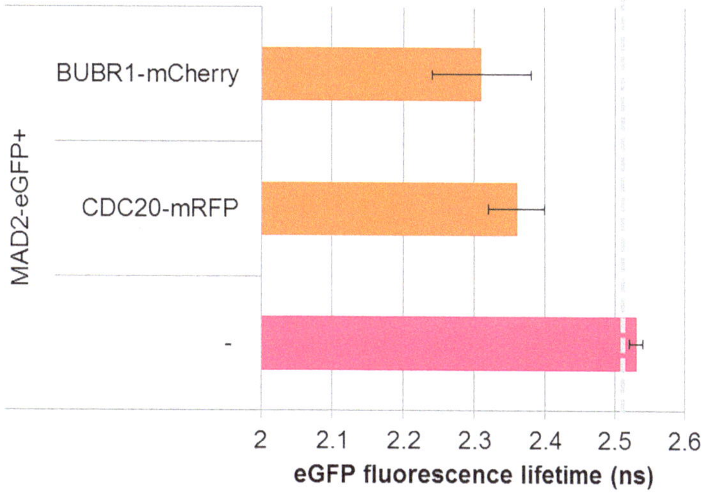

Fig. 3 Results of FRET-FLIM analysis of MAD2-eGFP protein-protein interactions. Bar graphs of the average fluorescence lifetime of MAD2-eGFP alone (−) or when co-expressed with CDC20-mRFP and BUBR1-mCherry. The gray dashed line represents the minimum fluorescence lifetime expected in the case of a non-interaction. Experimental pairs with eGFP lifetimes shorter than this are therefore interacting. Error bars represent the standard deviation of each experiment

4 Notes

1. Additional system requirements that will increase the speed of analysis: a memory of 512 Mb or more, 5Gb free space on the hard disk and a large, high quality monitor, 17″ or larger.
2. Cut tobacco leaf samples should measure approximately 5mm^2. The sections should be taken from an area of leaf that is not directly next to the site of infiltration as this area may have some necrosis. Leaf veins should also be avoided as they make mounting the leaf sample flare against the slide challenging.
3. It is important that the coverslip is properly secured. Sample drift during image acquisition will affect data quality.
4. Reducing background autofluorescence is essential, as it may affect the results obtained during FRET-FLIM acquisition and may obscure the emission of the chosen FRET pair.
5. A suitable negative control is a donor expressed with no acceptor, whilst for positive controls it is best to select a pair of proteins that are already known to interact.

6. The selected region of interest should be as close to the structure of interest as possible, taking care to exclude any unwanted fluorescence. For example, if there are chloroplasts visible in the image, they may have a distinct, detectable fluorescence lifetime. To ensure that the fluorescence lifetime of the chloroplasts is not included in any calculations, any chloroplasts should be outside the selected region of interest.
7. If most values have a poor fit, there may be issues with the experimental setup, such as insufficient donor expression, too high intensity of the selected fluorophores, too short an exposure time, or an inappropriate decay model.
8. These results can be presented as a bar graph as shown in Fig. 3. It is important to display the negative control of a single fluorophore alongside the tested interactions as though the FLIM of a fluorophore can be calculated in theory; in practice, FLIM is dependent on the microenvironment in which the fluorophore is present.

Acknowledgments

This work was supported by an Early Career Fellowship grant from the Leverhulme Trust (ECF-2022-002).

References

1. Komaki S, Schnittger A (2016) The spindle checkpoint in plants—a green variation over a conserved theme. Curr Opin Plant Biol 34:84–91
2. Sudakin V, Chan GK, Yen TJ (2001) Checkpoint inhibition of the APCC in HeLa cells is mediated by a complex of BUBRI, BUB3, CDC20, and MAD2. J Cell Biol 154:925–936
3. Herzog F, Primorac I, Dube P, Lénárt P, Sander B, Mechtler K et al (2009) Structure of the anaphase-promoting complex/cyclosome interacting with a mitotic checkpoint complex. Science 323:1477–1481
4. Li Y, Gorbea C, Mahaffey D, Rechsteiner M, Benezra R (1997) MAD2 associates with the cyclosome/anaphase-promoting complex and inhibits its activity. Proc Natl Acad Sci USA 94:12431–12436
5. Fang G, Yu H, Kirschner MW (1998) Direct binding of Cdc20 protein family members activates the anaphase-promoting complex in mitosis and G1. Mol Cell 2:163–171
6. Hwang LH, Lau LF, Smith DL, Mistrot CA, Hardwick KG, Hwang ES et al (1998) Budding yeast Cdc20: a target of the spindle checkpoint. Science 279:1041–1044
7. Kallio M, Weinstein J, Daum JR, Burke DJ, Gorbsky GJ (1998) Mammalian p55CDC mediates association of the spindle checkpoint protein Mad2 with the cyclosome/anaphase-promoting complex and is involved in regulating anaphase onset and late mitotic events. J Cell Sci 141:1393–1406
8. Kim SH, Lin DP, Matsumoto S, Kitazono A, Matsumoto T (1998) Fission yeast Slp1: an effector of the Mad2-dependent spindle checkpoint. Science 279:1045–1047
9. Wassmann K, Benezra R (1998) Mad2 transiently associates with an APC/p55Cdc complex during mitosis. Proc Natl Acad Sci USA 95:11193–11198
10. Brady DM, Hardwick KG (2000) Complex formation between Mad1p, Bub1p and Bub3p is crucial for spindle checkpoint function. Curr Biol 10:675–678
11. Taylor SS, Ha E, McKeon F (1998) The human homologue of Bub3 is required for kinetochore localization of Bub1 and a

Mad3/Bub1-related protein kinase. J Cell Biol 142:1–11

12. Logarinho E, Bousbaa H (2008) Kinetochore-microtubule interactions "in check" by Bub1, Bub3 and BubR1: the dual task of attaching and signalling. Cell Cycle 7(12):1763–1768
13. Kriechbaumer V, Botchway SW, Slade SE, Knox K, Frigerio L, Oparka K et al (2015) Reticulomics: protein-protein interaction studies with two Plasmodesmata-localized Reticulon family proteins identify binding partners enriched at Plasmodesmata, endoplasmic reticulum, and the plasma membrane. Plant Physiol 169(3):1933–1945
14. Margineanu A, Chan JJ, Kelly DJ, Warren SC, Flatters D, Kumar S et al (2016) Screening for protein-protein interactions using Förster resonance energy transfer (FRET) and fluorescence lifetime imaging microscopy (FLIM). Sci Rep 6(28186):1–15
15. Duckney P, Hussey PJ (2023) Investigating plant protein–protein interactions using FRET-FLIM with a focus on the actin cytoskeleton. In: Hussey PJ, Wang P (eds) The plant cytoskeleton, Methods in Molecular Biology. Humana, New York
16. Rios AF, Radoeva T, De Rybel B, Weijers D, Borst JW (2017) FRET-FLIM for visualizing and quantifying protein interactions in live plant cells. In: Kleine-Vehn J, Sauer M (eds) Plant hormones, Methods in Molecular Biology. Humana Press, New York
17. Förster T (1965) Delocalized excitation and excitation transfer. Modern Quantum Chem 3:93–137
18. Stryer L (1978) Fluorescence energy transfer as a spectroscopic ruler. Annu Rev Biochem 47:819–846
19. Wu P, Brand L (1994) Resonance energy transfer: methods and applications. Anal Biochem 218(1):1–13
20. Lakowicz JR (2013) Principles of fluorescence spectroscopy, 2nd edn. Springer, New York
21. Datta R, Heaster TM, Sharick JT, Gillette AA, Skalaa MC (2020) Fluorescence lifetime imaging microscopy: fundamentals and advances in instrumentation, analysis, and applications. J Biomed Opt 25(7):1–43
22. Koushik SV, Vogel SS (2008) Energy migration alters the fluorescence lifetime of cerulean: implications for fluorescence lifetime imaging Forster resonance energy transfer measurements. J Biomed Opt 13(3):1–18
23. Tadrous PJ (2000) Methods for imaging the structure and function of living tissues and cells: 2. Fluorescence lifetime imaging. J Pathol 191:229–234
24. Chen L-C, Lloyd WR 3rd, Chang C-W, Sud D, Mycek M-A (2013) Fluorescence lifetime imaging microscopy for quantitative biological imaging. Methods Cell Biol 114:457–488
25. Mastop M, Bindels DS, Shaner NC, Postma M, Gadella TWJ Jr, Goedhart J (2017) Characterization of a spectrally diverse set of fluorescent proteins as FRET acceptors for mTurquoise2. Sci Rep 7:11999
26. Sekar RB, Periasamy A (2003) Fluorescence resonance energy transfer (FRET) microscopy imaging of live cell protein localizations. J Cell Biol 160(5):629–633
27. Spatola Rossi T, Pain C, Botchway SW, Kriechbaumer V (2022) FRET-FLIM to determine protein interactions and membrane topology of enzyme complexes. Current Protocols 2:e598
28. Elangovan M, Day RN, Periasamy A (2002) Nanosecond fluorescence resonance energy transfer-fluorescence lifetime imaging microscopy to localize the protein interactions in a single living cell. J Microsc 205:3–14
29. Lambert TJ (2019) FPbase: a community-editable fluorescent protein database. Nat Methods 16:277–278
30. Long Y, Stahl Y, Weidtkamp-Peters S, Smet W, Du Y, Gadella TWJ Jr et al (2018) Optimizing FRET-FLIM labeling conditions to detect nuclear protein interactions at native expression levels in living Arabidopsis roots. Front Plant Sci 9(369):1–13
31. Hecker A, Wallmeroth N, Peter S, Blatt MR, Harter K, Grefen C (2015) Binary 2in1 vectors improve in planta (co)localization and dynamic protein interaction studies. Plant Physiol 168(3):776–787
32. Denay G, Schultz P, Hänsch S, Weidtkamp-Peters S, Simon R (2019) Over the rainbow: a practical guide for fluorescent protein selection in plant FRET experiments. Plant Direct 3(12):e00189

Chapter 9

Using FRET to Define Cdk1-Dependent Ordering of Events During Exit from Second Meiotic M-Phase in Oocytes

Chenxi Zhou, Yunan Ye, and Hayden Homer

Abstract

Exit from M-phase requires a precise sequence of molecular events for successful completion, with errors in the process resulting in cell death or aneuploidy, a characteristic feature of cancer and the leading cause of pregnancy failure. Exit from the second meiotic division (MII) in oocytes is a unique event triggered by sperm, involving female anaphase II as well as both male and female pronuclear formation. Very little is known about how these events involving two distinct cell types are coordinated. M-phase exit is driven by inactivation of the master cell-cycle regulator, cyclin-dependent kinase 1 (Cdk1), but details of how Cdk1 orchestrates MII exit has remained sketchy due to technical challenges in studying these events. Here we detail a protocol for undertaking in-depth analysis of Cdk1 activity throughout fertilization in live mouse oocytes using a Cdk1 Fluorescence Resonance Energy Transfer (FRET) biosensor. This protocol illustrates the utility of time-lapse imaging and FRET for interrogating experimentally challenging cell-cycle events.

Key words Cdk1, FRET, Meiosis, Fertilization, M-phase exit, Pronuclear formation

1 Introduction

Exit from M-phase of the cell cycle, which has been most extensively studied during mitosis, involves chromosome segregation (anaphase), cytokinesis/abscission, chromosome decondensation and nuclear reformation [1, 2]. Stepwise inactivation of the master cell-cycle regulator, cyclin-dependent kinase 1 (Cdk1), a heterodimer composed of a catalytic Cdk1 subunit and a cyclin B1 co-activator, and ordered dephosphorylation of Cdk1 substrates is crucial for orchestrating mitotic exit such that anaphase precedes cytokinesis and so on [1, 2].

Fluorescence resonance energy transfer (FRET) is a process whereby energy is transferred between two fluorescent molecules that are very close to one another, typically within 10 nm [3, 4]. This transfer occurs when one molecule (the donor) is excited by a light source and passes its energy to the other molecule

Victor M. Bolanos-Garcia (ed.), *Mitotic Exit: Methods and Protocols*, Methods in Molecular Biology, vol. 2874, https://doi.org/10.1007/978-1-0716-4236-8_9,

(the acceptor), causing it to emit light. The use of FRET biosensors in cell biology has revolutionized our understanding of complex biological processes [5]. By providing real-time monitoring of molecular interactions, FRET biosensors have enabled researchers to study cellular processes in unprecedented detail [6].

Using a Cdk1 FRET biosensor, it was shown that sequential steps during entry to mitosis depend on increasing levels of Cdk1 activity [7]. For this biosensor, the donor and acceptor fluorescent proteins are linked by a Cdk1-specific phosphorylation site (16 amino acids from cyclin B1 encompassing a Cdk1 autophosphorylation site) and a phosphobinding domain (the polo box domain [PBD] of Polo-like kinase 1 [Plk1]) that binds to Cdk1 phosphorylation sites (Fig. 1a). Phosphorylation of the linker site results in its binding to the phosphobinding domain, thereby inducing a conformational change that approximates the donor and acceptor resulting in FRET (Fig. 1).

Meiosis involves two sequential M-phases without an intervening S-phase thereby halving the chromosome complement. For mammalian oocytes, the second meiotic division (MII) progresses to metaphase II at which point oocytes arrest under the influence of cytostatic factor with a fully formed bipolar spindle sustained by high Cdk1 activity. MII arrest is broken when Cdk1 inactivation is triggered by sperm entry at fertilization following which, oocytes undergo exit from MII involving anaphase II, cytokinesis (resulting in extrusion of the second polar body [PB]) and formation of both the female and male pronuclei (PN). Notably, unlike minutes-long mitotic exit, exit from female MII is an hours-long process important for male PN formation [8]. Due to the requirement for fertilization and its protracted duration, details of exit from MII and the role of Cdk1 inactivation have been technically challenging to investigate.

Here we present a protocol involving the Cdk1 FRET biosensor and time-lapse imaging of chromosomes and spindles in live mouse oocytes undergoing fertilization to study how Cdk1 regulates female MII exit events. Using this protocol, we recently discovered that during exit from female MII, Cdk1 inactivation occurs over two phases separated by a 3-h hiatus and that this hiatus of sustained low-level Cdk1 activity is pivotal for sperm chromatin reconfiguration needed for male PN formation [9].

2 Materials

2.1 Mouse Oocyte Collection and Culture

1. Stereo microscope (e.g., Leica MDG 30).
2. Fine forceps.
3. Scissors.
4. 35 × 10 mm polystyrene Petri dishes.

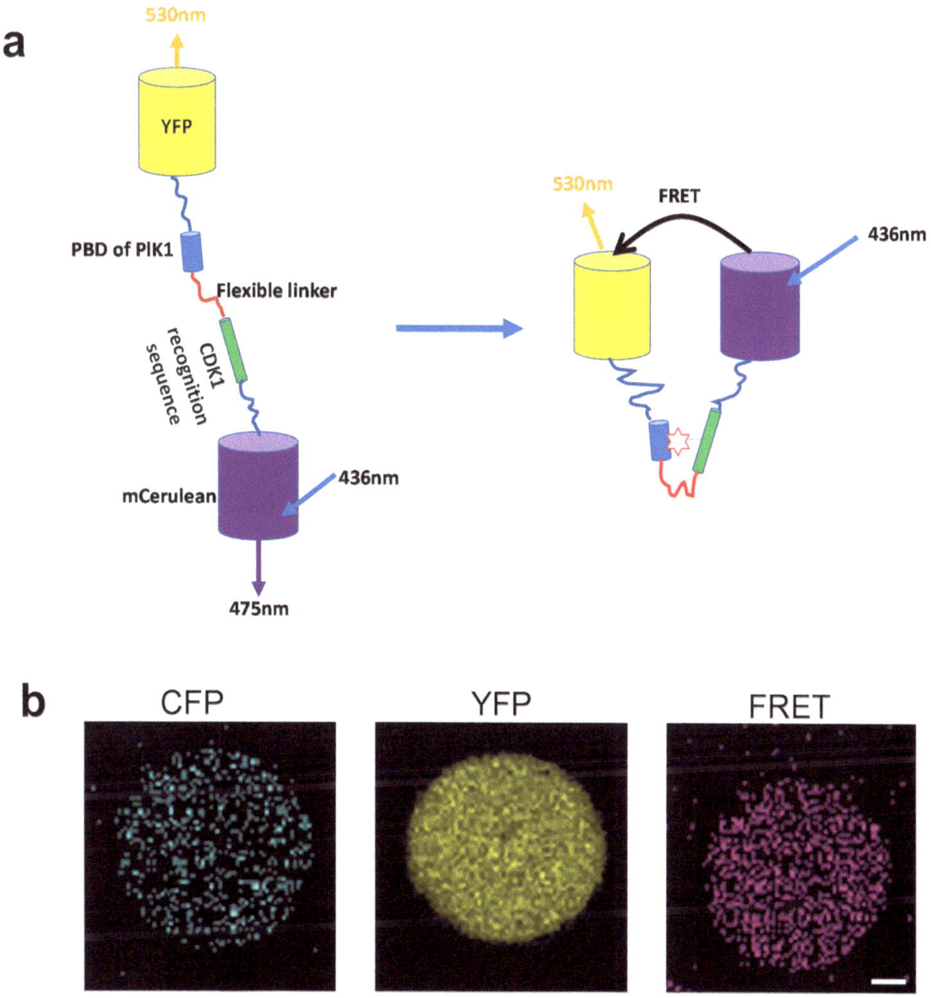

Fig. 1 (**a**) Schematic of the Cdk1 biosensor. The biosensor comprises the FRET pair (YFP and mCerulean), the Polo-Box Domain of Plk1 (residues 373–592, blue), a central flexible linker, and a Cyclin B1 peptide containing the Ser126 autophosphorylation site (green). (**b**) Representative fluorescence images of oocyte expressing CFP, YFP, and the FRET probe. Scale bars: 10 μm

5. M2 and M16 medium.
6. Microcentrifuge tubes or Eppendorf tubes.
7. Embryo-tested light mineral oil.
8. CO_2 incubator.
9. Mouth pipettes. The mouth pipette assembly consists of a hard plastic mouthpiece connected by latex tubing to a flame-pulled glass pipette (*see* **Note 1**) via a 1000 μL plastic pipette tip; the broad end of the glass pipette is inserted into the broad end of the pipette tip while the tip is inserted into the latex tubing.

2.2 Oocyte Microinjection

1. Microinjection microscope.
2. Holding pipettes (inner diameter 15 μM, outer diameter 75 μM, 35° bend; The Pipette Company).
3. Capillary tubes (0.86 mm inner diameter, 1.5 mm outer diameter).
4. Embryo-tested light mineral oil.
5. M2 and M16 medium.
6. In vitro transcribed complementary (cRNA) for expression.
7. Sterile pipettes and tips (for handling cRNA).
8. Eppendorf tubes (for preparing injection solutions).
9. CO_2 incubator (for maintaining injected oocytes in culture).
10. 35 × 10 mm polystyrene Petri dishes.
11. Microinjection pipette puller (P30 vertical micropipette puller, Sutter Instruments).
12. Glass capillaries with filament (e.g., borosilicate glass, 1.0 mm diameter, 10 cm length).
13. Microinjection equipment: Intracellular electrometer (IE-251A, Warner Instruments and Pneumatic PicoPump (PV-820, World Precision Instruments).

2.3 In Vitro Fertilization

1. Dissecting microscope with a heated stage (Tokai Hit).
2. Fine forceps.
3. Scissors.
4. M2 medium.
5. Human Tubal Fluid medium (HTF) (*see* **Note 2**).
6. Mineral oil.
7. CO_2 incubator.
8. Sterile pipettes.
9. KSOM medium (EmbryoMax® Advanced KSOM Embryo Medium).
10. Culture dishes.

2.4 cRNA Preparation

1. DNA plasmids for Cdk1 FRET probe (Fig. 1a), GFP, YFP, H2B-RFP.
2. Plasmid extraction kit (miniprep).
3. Restriction enzyme and buffer (for DNA linearization).
4. Phenol:Chloroform:IAA, 25:24:1.
5. Pellet paint.
6. 3 M Sodium acetate solution, pH 5.2.
7. Chloroform.

8. Proteinase K.
9. 10% SDS.
10. 100% Ethanol.
11. T7, T3, or SP6 in vitro transcription kit.
12. Nuclease-free water (for RNA dilution and storage).
13. UV spectrophotometer or nanodrop (for RNA quantification).
14. Nuclease-free microcentrifuge tubes or Eppendorf tubes (for eluting cRNA).
15. Benchtop centrifuge (for RNA precipitation and pelleting).
16. Heat block or thermocycler (for incubating reactions at appropriate temperatures).
17. Nuclease-free pipettes, tips, and tubes (for handling RNA).
18. −20 °C and − 80 °C freezers.

2.5 Confocal Microscopy and Cdk1 Measurement

1. Confocal microscope (Leica TCS SP8 confocal microscope).
2. Nunc™ Glass Bottom Dishes (Thermo Scientific).
3. Live-cell stage-mounted imaging chamber for maintaining ambient conditions suitable for fertilization.
4. A wide-field fluorescence inverted microscope equipped with a 20 × /0.7 NA objective.
5. Excitation laser(s).
6. Computer and software (for automated control of image acquisition).
7. Image analysis software (for processing and analyzing acquired images, such as Image J, LAS X from Leica).
8. Live cell imaging medium (for imaging fertilization process, use MEMα without BSA).
9. Sterile pipettes and tips.

3 Methods

3.1 Mouse Oocyte Collection and Culture

1. Obtain female mice that are 3–4 weeks old (*see* **Note 3**).
2. Inject 5–10 international units (IU) of pregnant mare's serum gonadotrophin PMSG intraperitoneally (IP) into each mouse. The timing of the injection will depend on the specific experimental protocol, but typically it is given in the afternoon (*see* **Note 4**).
3. Pre-warm an aliquot of M2 supplemented with 50 μM 3-isobutyl-1-methylxanthine (IBMX) to prevent oocytes from resuming meiotic maturation on a 37 °C heat block for about 20 minutes.

4. Sacrifice mice by cervical dislocation or other approved method 44–48 h after PMSG injection.
5. Using scissors and forceps, dissect the female and rapidly collect both ovaries in 2 mL of pre-warmed M2 medium supplemented with 50 μM IBMX in a watch glass dish on a 37 °C heating plate and transfer to the 37 °C heated stage of the dissection scope.
6. Use fine forceps to stabilize the ovary and 27 G needles to puncture the ovarian follicles and release the oocytes into a 35 × 10 mm dish containing M2 + IBMX medium.
7. Using a stereomicroscope (e.g., a M165C, Leica Microsystems), identify the immature oocytes by the presence of a large, round nucleus known as a germinal vesicle (GV) [10].
8. Using the mouth pipette, gently aspirate and expel the oocytes through the fine tip of a glass pipette, repeating as many times as needed, in order to remove the covering cumulus cells (known as denudation) surrounding the oocytes (*see* **Note 1**). Only fully-grown cumulus-covered oocytes should be isolated and used for experiments as these are the most developmentally competent (*see* **Note 4**).
9. Transfer denuded oocytes to clean M2 with IBMX drops.
10. Allow oocytes to recover for about 20 minutes on the 37 °C heat stage before proceeding to cRNA injection.

3.2 cRNA Preparation

1. Ensure correct plasmid identity by sequencing and grow bacterial cultures containing the correct plasmid.
2. Extract template DNA from the bacteria using a standard plasmid miniprep procedure and determine its concentration using a spectrophotometer.
3. Based on the concentration obtained, calculate the volume (x) of plasmid DNA required to achieve a range of 2–4 μg. Linearize the plasmid DNA using a restriction enzyme targeting downstream of the sequence of interest by incubating the following mixture for >5 h. Calculate the volume (y) of nuclease-free water by subtracting the sum of the first three components from the total volume of 100 μL (*see* **Note 5**).

x μL	Plasmid DNA (2–4 μg)
2 μL	Restriction enzyme
10 μL	10x reaction buffer
y μL	Nuclease-free water
100 μL	

4. Remove RNase by adding 1 μL of Proteinase K (10 μg/ml), 5.3 μL 10% SDS and 50 μL nuclease-free water and incubating at 50 °C for 1 h.
5. Add 150 μL phenol:chloroform:IAA, 25:24:1, pH 7.9, vortex vigorously to mix and centrifuge at maximum speed for 10 min.
6. Carefully transfer the upper aqueous layer to a new RNase-free tube. Add 150 μL chloroform, mix and centrifuge for another 10 min.
7. Transfer the upper aqueous layer to a new RNase-free tube and add 10% volume RNase-free 3 M Sodium Acetate (e.g., 15 μL for 150 μL chloroform extract), 2 μL pellet paint, and 450 μL 100% ethanol. Incubate the mixture at −80 °C for 2 h.
8. Centrifuge at maximum speed for 20 minutes and discard the supernatant.
9. Wash the pellet with 500 μL 70% ethanol (freshly prepared using RNase-free water).
10. Remove all traces of ethanol by centrifugation and air dry the pellet.
11. Resuspend the template DNA in 6 μL RNase-free water. This can be used immediately for in vitro transcription or stored at −20 °C.
12. Assemble the transcription reaction according to kit procedure. The following mixture describes in vitro transcription using a T7 mMESSAGE mMACHINE Kit. Calculate the volume (y) of nuclease-free water by subtracting the sum of the first four components from the total volume of 20 μL.
13. Incubate the mixture at 37 °C for >2 h. For transcripts longer than 5 kb, a longer incubation time is recommended.
14. Add 1 μL DNase into the mixture, mix thoroughly and incubate 37 °C for 15 minutes to remove the template DNA.
15. Recover RNA by lithium chloride (LiCl) precipitation. Add 30 μL nuclease-free water and 30 μL LiCl precipitation solution. Incubate at −20 °C for >30 minutes or overnight.
16. Centrifuge at maximum speed in a 4 °C centrifuge to pellet the RNA.
17. Remove the supernatant on ice and wash with 1 mL 70% ethanol.
18. Centrifuge again to remove all traces of ethanol.
19. Resuspend the dried RNA pellet in 20 μL nuclease-free water.
20. Determine the concentration of cRNA using a spectrophotometer and make 1 μL aliquots in nuclease-free

microcentrifuge tubes. Freeze the aliquots at −80 °C until microinjection.

6 μL	Linear DNA template
10 μL	2X NTP/CAP
2 μL	10X reaction buffer
2 μL	Enzyme mix
Y μL	Nuclease-free water
20 μL	

3.3 Oocyte Microinjection

1. Prior to microinjection, thaw an aliquot of the cRNA solution on ice, centrifuge briefly and dilute to the desired concentration. As a guide, we use H2B-RFP at 250 ng/μL, CFP at 200 ng/μL, YFP at 200 ng/μL and Cdk1 FRET probe at 500 ng/μL.
2. Prepare a microinjection needle using a capillary tube (0.86 mm inner diameter, 1.5 mm outer diameter) and vertical pipette puller (*see* **Note 6**).
3. Load cRNA solution into the tip of microinjection needle and remove air bubbles by gently tapping the needle.
4. Insert the microinjection needle and the holding pipette into the microinjection and holding arms, respectively, of the micromanipulation assembly mounted on the microscope.
5. Prepare a microinjection dish by pipetting 1 mL of IBMX-treated M2 medium in to a 35 mm culture dish and cover with mineral oil.
6. Transfer denuded oocytes into the medium in the dish and place on the stage of the microinjection microscope.
7. Using the micromanipulators, carefully lower the holding and microinjection pipettes into the medium so that they are in the same horizontal plane as the oocytes.
8. Stabilize a single oocyte by applying suction to the tip of the holding pipette using a manual syringe device.
9. Using the micromanipulator, advance the tip of the microinjection needle across the zona pellucida and oolemma of the stabilized oocyte so that it is seen to enter the ooplasm.
10. Apply a brief pulse of negative capacitance via an intracellular electrometer.
11. Expel a controlled volume of cRNA solution (equivalent to ~5% of the oocyte volume) using a device such as a Pneumatic PicoPump.

12. Following microinjection, transfer oocytes to IBMX-treated M16 medium and culture for >2 h to allow time for protein translation.
13. Wash oocytes through 5–6 drops of IBMX-free M16 medium to initiate resumption of meiotic maturation and culture overnight.
14. On the following morning, isolate MII-arrested oocytes (identifiable by the presence of a PB) with adequate fluorescence (indicating robust expression of injected cRNA constructs; *see* Fig. 1b) for subsequent in vitro fertilization.

3.4 In Vitro Fertilization (IVF)

1. Prepare sperm dishes. Put 1 drop (100 μL / drop) of HTF medium (*see* **Materials 2.3.5**) into a dish and cover it with mineral oil (*see* **Materials 2.3.6**) 30 min before collecting sperm, and place the dish in a 37 °C incubator.
2. Prepare fertilization dish. Put 1 drop (200 μL / drop) of HTF medium into a dish and cover it with mineral oil 30 min before fertilization, and place the dish in an incubator.
3. Prepare washing dishes. Put 3 drops (50 μL / drop) of MEMα without BSA medium (*see* **Materials 2.5.9**) into a Petri dish and cover it with mineral oil 30 minutes before fertilization.
4. Prepare live image dishes. Put one 5 μL drop of MEMα without BSA medium into Scientific™ Nunc Glass Bottom Dish (*see* **Materials 2.5.2**), cover with mineral oil and place in a 37 °C incubator.
5. Obtain male mice that are at least 12–16 weeks old (*see* **Note** 7).
6. Sacrifice a male mouse by cervical dislocation.
7. Use scissors to make a small incision on the lower abdomen and expose the testes.
8. Gently exteriorize each testis, dissect off the cauda epididymis and place it in a dish containing pre-warmed HTF medium to prevent dehydration.
9. Collect sperm by placing the epididymis into pre-warmed HTF medium (*see*, **step 3.4.1**) and using scissors or fine forceps to make a small incision in the capsule of each epididymis. Leave for 37 °C for 15–30 min to allow the sperm to swim out of the tissue.
10. Remove the epididymis, leave the sperm in culture for a further 3 h to enable capacitation.
11. Evaluate sperm motility and estimate concentration using a stereomicroscope (*see* **Note 8**).
12. Transfer fluorescent MII oocytes from **step 3.3.14** to a droplet of acidic Tyrode's solution in a culture dish (without overlaying

oil) and incubate for 20–60 s with gentle agitation (*see* **Note 9**).

13. After the incubation period, quickly remove oocytes from Tyrode's solution and transfer to a fresh droplet of untreated medium (*see* **Note 9**).
14. Wash the oocytes through several droplets of untreated medium to ensure complete removal of the acidic Tyrode's solution.
15. Transfer zona-free MII oocytes into fertilization dishes (*see* **step 3.4.2**) and then incubate in an incubator (37 °C, 5% CO_2 in air) for 30 minutes before insemination.
16. Use the tip of a pipette to add appropriate amounts (usually about 3 μL) of the sperm suspension to the fertilization drop of HTF containing the zona-free MII oocytes from, **step 3.4.13**.
17. Place the fertilization dish with both sperm and oocytes in an incubator (37 °C, 5% CO_2 in air) for 15 min (*see* **Note 8**).
18. After 15 minutes co-incubation, transfer the zona-free MII oocytes attached with sperm from the HTF fertilization medium to one of the drops of pre-warmed MEMα without BSA medium (*see* **step 3.4.3**) by mouth pipetting. Thoroughly remove unattached sperm from contact with oocytes by sequentially transferring them through 3 drops of MEMα without BSA medium (*see* **step 3.4.3**, *see* **Notes 10** and **11**).
19. Transfer inseminated oocytes (with attached sperm) to an imaging dish from **step 3.4.4** by mouth pipetting and perform live-cell confocal imaging for 15 h to capture the entire process of fertilization and exit from MII (*see* **Notes 10** and **11**).

3.5 Measuring Cdk1 Activity Using FRET

1. Place the dish containing inseminated oocytes (with attached sperm from **step 3.4.19**) on the microscope stage. We use a Leica TCS SP8 confocal microscope equipped with a 20x/0.75 NA Apochromat water-immersion objective and an automated pump cap (water-immersion micro dispenser, Leica; automated pump mp-x controller, Bartels Mikrotechnik) [9–12]. Stable environmental conditions of 37 °C and 5% CO_2 are maintained throughout the entire experimental procedure using a stage-mounted chamber enclosed within a custom-designed polycarbonate incubator (Life Imaging Services).
2. Using Leica LAS X software, select the appropriate lasers and filter sets for imaging different combinations of Cdk1 FRET, H2B-RFP, YFP, and CFP as required (Fig. 2a, b). For H2B-RFP use a 546 ± 6 nm excitation filter, a 565 dichroic mirror and a 620 ± 20 nm emission filter (Fig. 2b); for CFP use a 440AF21 excitation filter, a 455 dichroic mirror, and 480AF30 emission filter; for YFP use 500AF25 excitation,

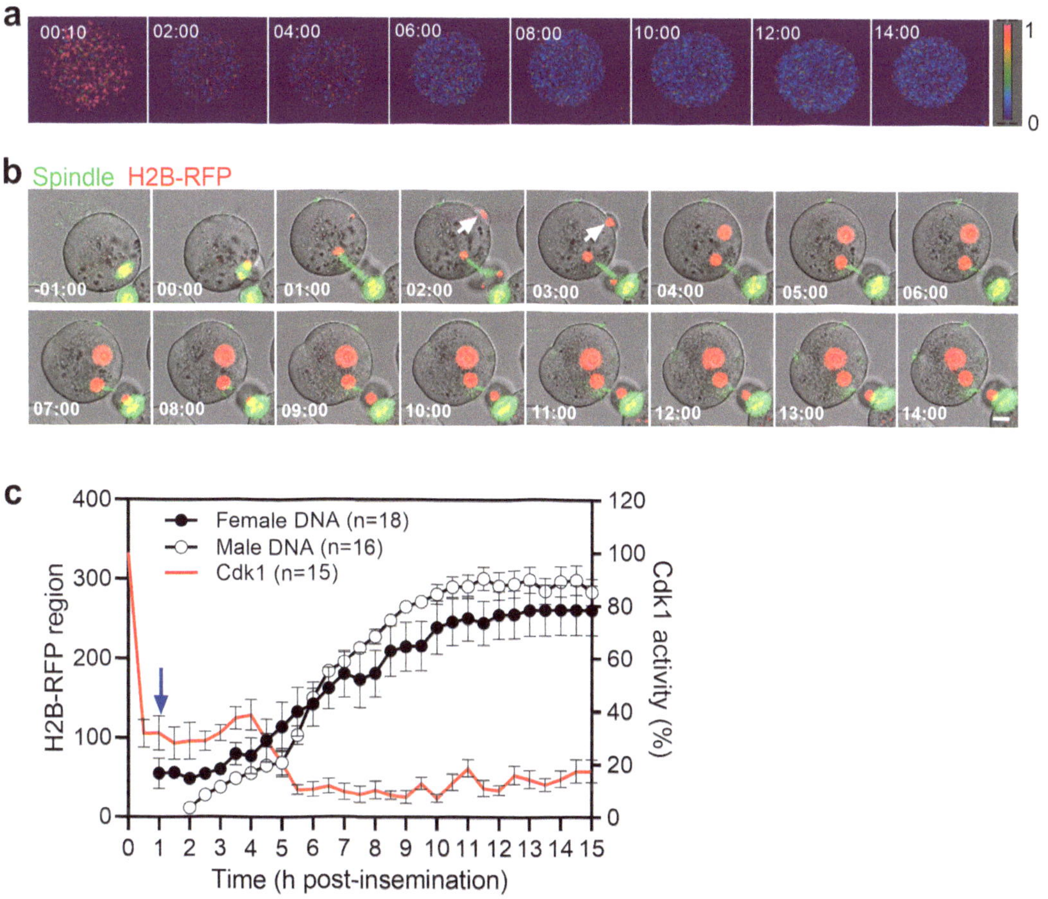

Fig. 2 (**a**) Representative time-lapse images of FRET changes following insemination of MII oocytes. Time is hh:mm post-insemination. Scale bars: 10 μm. (**b**) Representative time-lapse images of spindles and chromosomes in live MII oocytes undergoing fertilization. White arrow, fertilization cone, and sperm DNA. Time is hh:mm post-anaphase-onset. Scale bar, 10 μm. (**c**) FRET ratios normalized to 100% at the time of insemination and sizes of the H2B-RFP signals for male and female DNA in live MII oocytes during fertilization. FRET ratio at the time of insemination ranged from 0.78 to 0.81. Blue arrow highlights time of anaphase II. Data are Mean ± SEM

545AF35 emission, and 525DRLB dichroic mirror; and for FRET use 440AF21 excitation, 535AF26 emission and a 455DRPL dichroic mirror.

3. Locate the oocytes in the z-axis and derive a stack thickness of 50 μm incorporating the oocytes' position. Acquire images at step sizes of 4 μm in the z-axis, using a speed of 600 Hz and at 10 min-intervals (Fig. 2a, c, *see* **Note 12**).
4. Activate the Leica FRET Sensitized Emission (SE) wizard located within the Leica Microsystems LASAF drop-down menu (Fig. 3c **1**).

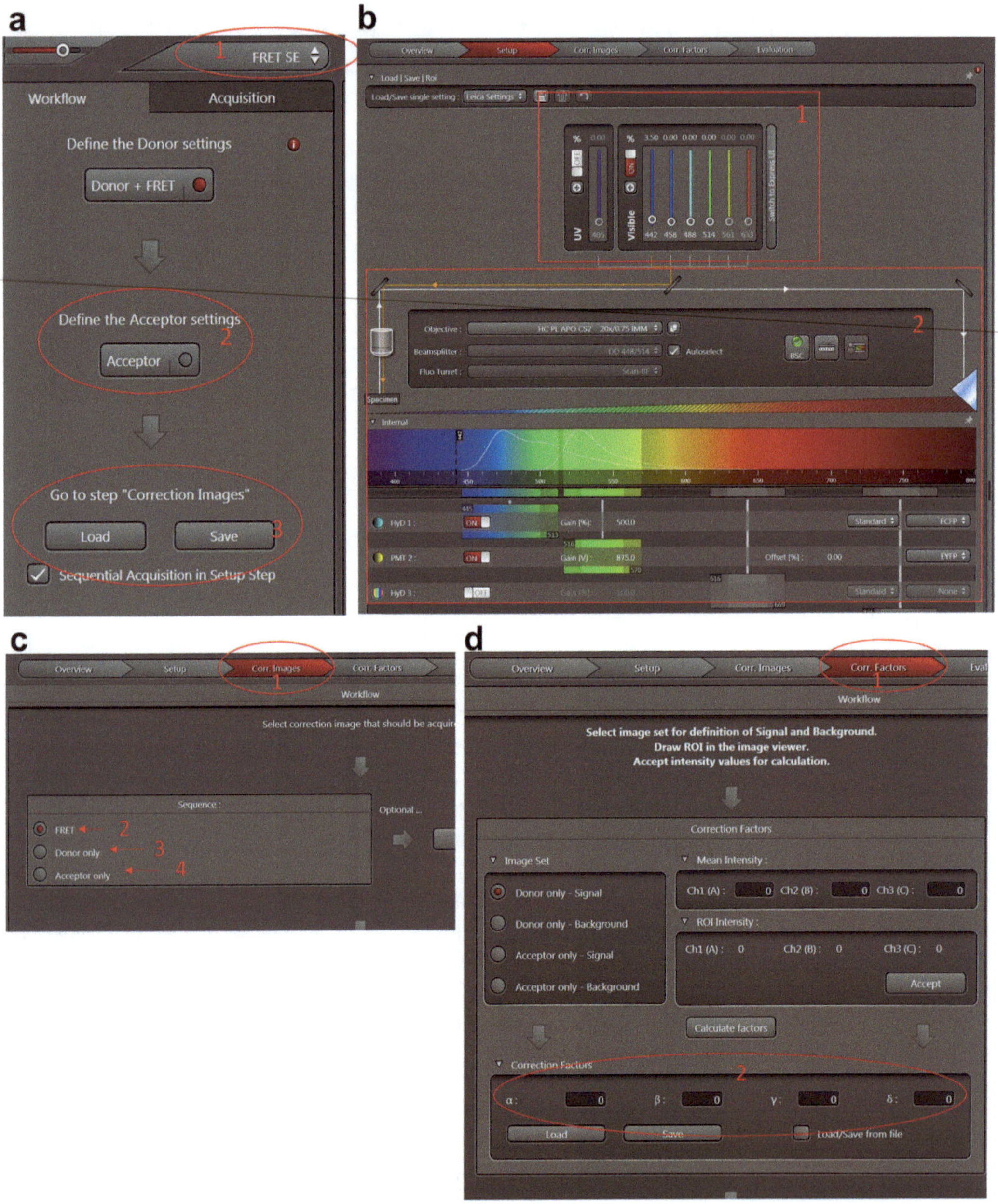

Fig. 3 Screen snapshots of Leica LASX software illustrating the setup process for a FRET experiment. (**a-b**) Software menu for activating the FRET SE wizard and acquiring images for Donor + FRET and Acceptor. (**c-d**) Software menu for capturing correction images and obtaining the Correction Factors

5. Activate the laser lines corresponding to Donor (476 nm laser for CFP) and Acceptor (514 nm laser line for YFP) excitation and simultaneously excite both (Fig. 3b **1**). Using the "Live" button and the tools menu under the "Acquisition" tab, make adjustments to obtain appropriate image resolution (*see* **Note 13**).

6. For Acceptor Setup, select the "Acceptor" table (514 nm laser for YFP) (Fig. 3a **2,** b **1–2**). Slowly increase the excitation power until the signal is just below saturation. Do NOT change the PMT gain or offset, as this will also change the imaging conditions for the FRET signal detection. Define the number of averages for best imaging conditions using the tools menu under the "Acquisition" tab.
7. Obtain Correction Images using the following steps (Fig. 3a **3,** c **1**). Acquire a FRET sequence (of the FRET sample, which was previously positioned on the microscope) using "Capture Image" and selecting the "FRET" button (Fig. 3c **2**). Place the Donor sample on the stage and acquire the Donor sequence using "Capture Image" and selecting the "Donor Only" button (Fig. 3c **3**). Place the Acceptor sample on the stage and acquire the Acceptor sequence using "Capture Image" and selecting the "Acceptor Only" button (Fig. 3c **4**). The acquired images will automatically be named "FRET," "Donor Only," and "Acceptor Only," respectively. Finally, use the tab labeled "Correction Factors" (Fig. 3d **1–2**) to generate the calibration factors.
8. Acquire timelapse FRET images of all samples in sequential mode (line by line) (*see* **Note 14**).
9. Once the FRET images of all samples have been collected, calculate FRET ratios using the Leica FRET SE Wizard, which is based on the formula described by Wouters and collaborators [13]:

$$E_A(I) = \frac{B–A \times b–C \times c}{C}$$

$EA(i)$ is the FRET efficiency.

$A \times b$ is the Donor cross-talk correction.

$C \times c$ is the Acceptor cross-excitation correction.

FRET $= B$.

Donor emission $= A$.

Donor emission+ cross-talk ration $= b$.

Acceptor emission $= C$.

Acceptor excitation cross-talk ration = c (*see* **Note 15**).

10. Normalize FRET ratios using values at the time of insemination as 100% since this is when Cdk1 activity is maximal prior to fertilization-induced oocyte activation. Normalized data facilitate comparison between different experimental conditions (Fig. 2c).
11. Export the calculated results displayed in the "Statistics" tab as excel files.

4 Notes

1. To create a narrow-bore glass pipette for mouth pipetting, first hold the two ends of the glass pipette between the thumb and forefinger of each hand. Then, heat the pipette in the flame of a Bunsen burner until the glass begins to melt. Next, remove the pipette from the flame and quickly draw the glass out horizontally by pulling the hands apart. Break off the distal portion of the pipette to leave a suitable working length. The aim is to create a narrow-bore pipette that is slightly larger than the object being pipetted, such as an oocyte with a diameter of approximately 100 μm. For a visual representation, see Greaney and collaborators [10].
2. The fertilization medium for mouse IVF typically consists of a buffered salt solution that mimics the conditions found in the female reproductive tract. We use Human Tubal Fluid medium (HTF) supplemented with 4 mg/mL BSA to the medium. The medium should be made fresh on the day of the experiment and kept at 37 °C until use.
3. We favor oocytes from B6CBAF1 mice, which are the F1 offspring resulting from crossing female C57BL/6 J mice with male CBA/CaCrl mice. 3–4-week old females from this strain produce superior oocyte yields having favorable maturation characteristics compared to other strains.
4. It is important to inject PMSG and harvest ovaries at the appropriate time post-PMSG to ensure optimal time for proper follicular development. When collecting ovaries, take care in handling the ovaries to avoid damaging the follicles or oocytes. Keep ovaries in warm medium when transferring from animal room to the laboratory. Use fine forceps to gently puncture the follicles and release the oocytes.
5. For linearization of the DNA template, it is important to choose an appropriate restriction enzyme and to use its optimal digestion conditions. It is recommended to examine a fraction of the linearized template DNA on a gel as initiation of transcription is the rate-limiting step in in vitro transcription. The presence of more than one band indicates incomplete linearization.
6. We use a vertical pipette puller (P30 vertical micropipette puller, Sutter Instruments). The optimal parameters for pulling microinjection needles are determined empirically. The needle tip should be fine enough to not cause undue oocyte trauma, but of sufficient bore to ensure that cRNA solution can be delivered easily without becoming blocked. The rate of oocyte death following microinjection should be <10%.

7. To ensure the availability of high-quality sperm capable of fertilizing oocytes, it is necessary to collect sperm from male mice that are at least 12–16 weeks old. Mice younger than this age tend to produce sperm with lower numbers and reduced motility, which can result in experimental failures. Therefore, we specifically select male mice within this age range for sperm collection.
8. The HTF medium should be prepared freshly. Sperm capacitation is a crucial step for successful fertilization. Ensure that the sperm have been properly capacitated before adding them to the oocytes. For IVF, mouse sperm is used at a concentration of 1–five million sperm per ml of fertilization medium. It is important to note that the concentration of the sperm can affect not only the fertilization rate, but also the quality of live imaging. Too high a concentration of sperm can compromise image quality by inducing excessive movement of the oocytes. On the other hand, a low sperm concentration may lead to reduced fertilization rates.
9. Acidic Tyrode's solution is toxic to the oocytes, so it is important to monitor oocytes closely during the removal process and limit their exposure to the minimum time required for zona pellucida removal.
10. BSA-free medium is crucial for successful imaging experiments involving inseminated oocytes due to vigorous movement caused by sperm which can compromise image quality. Only in a BSA-free environment can the oocyte properly attach to the glass bottom of the dish to limit movement.
11. ZP-free oocytes tend to be sticky and can easily adhere to the bottom of the dish or the glass pipette during transfer. To minimize oocytes sticking to the pipette, we transfer only 1–2 oocytes at a time.
12. The levels of FRET, CFP, and YFP expression are crucial for the success of this experiment, as higher expression levels result in improved imaging and more precise calculation of FRET ratios. When CFP and YFP expression levels are too low, the correction parameters used for FRET calculation can be significantly affected. Therefore, microinjection of cRNA in higher concentrations is critical to achieving optimal expression levels of CFP, YFP, and FRET.
13. From this time on, laser radiation may be present in the specimen area of the laser scanning microscope. Make sure to follow the safety notes for operation of the system.
14. During the acquisition of the FRET signal, it is essential to maintain the measuring parameters such as gain, detection window, and excitation wavelength. consistent throughout the FRET, Donor-only, and Acceptor-only image acquisition

processes. Otherwise, the FRET signal may fail due to variations in the signal-to-noise ratio. This can lead to inaccurate or unreliable results, which may affect the interpretation of the data.

15. There are three calculation methods available to choose from in the FRET SE program with Method 1 being the default choice. The calculation method can be changed after completion of the experiment.

Acknowledgments

This work was supported by the Professor Christopher Chen Endowment Fund, startup funding from the Faculty of Medicine, University of Queensland, and National Health and Medical Research Council Project Grants APP1078134, APP1103689, and APP1122484 to H.A.H.

Y.Y. is supported by the University of Queensland Research Training scholarship.

This protocol was adapted with minor modification from a previous study we published [9].

References

1. Sullivan M, Morgan DO (2007) Finishing mitosis, one step at a time. Nat Rev Mol Cell Biol 8(11):894–903
2. Wolf F, Sigl R, Geley S (2007) The end of the beginning: cdk1 thresholds and exit from mitosis. Cell Cycle 6:1407–1410
3. Miyawaki A (2003) Visualization of the spatial and temporal dynamics of intracellular signaling. Dev Cell 4(3):295–305
4. Jares-Erijman EA, Jovin TM (2003) FRET imaging. Nat Biotechnol 21(11):1387–1395
5. Sako Y, Yanagida T (2003) Single-molecule visualization in cell biology. Nat Rev Mol Cell Biol 4(9):SS1–SS5
6. Hochreiter B, Garcia AP, Schmid JA (2015) Fluorescent proteins as genetically encoded FRET biosensors in life sciences. Sensors 15(10):26281–26314
7. Gavet O, Pines J (2010) Progressive activation of CyclinB1-Cdk1 coordinates entry to mitosis. Dev Cell 18(4):533–543
8. Soeda S, Yamada-Nomoto K, Michiue T, Ohsugi M (2018) RSKMASTL pathway delays meiotic exit in mouse zygotes to ensure paternal chromosome stability. Dev Cell 47: 363–376
9. Zhou C, Homer HA (2022) The oocyte spindle midzone pauses Cdk1 inactivation during fertilization to enable male pronuclear formation and embryo development. Cell Rep 39(5): 110789
10. Greaney J, Subramanian GN, Ye Y, Homer HA (2021) Isolation and in vitro culture of mouse oocytes. Bio Protocol. https://doi.org/10.21769/BioProtoc.4104
11. Zhou C, Hancock JL, Khanna KK, Homer HA (2019) First meiotic anaphase requires Cep55-dependent inhibitory cyclin-dependent kinase 1 phosphorylation. J Cell Sci 132(18): jcs233379. https://doi.org/10.1242/jcs.233379
12. Wei Z, Greaney J, Loh WN, Homer HA (2020) Nampt-mediated spindle sizing secures a post-anaphase increase in spindle speed required for extreme asymmetry. Nat Commun 11(1):3393. https://doi.org/10.1038/s41467-020-17088-6
13. Wouters FS et al (2001) Imaging biochemistry inside cells. Trends Cell Biol 11:203–211

Chapter 10

Time-Lapse Imaging to Analyze Cell Fate in Response to Antimitotics

Patrícia M. A. Silva, João P. N. Silva, Bárbara Pinto, and Hassan Bousbaa

Abstract

Time-lapse imaging is a powerful technique widely used in Cellular and Molecular Biology to capture and analyze the dynamic processes of living cells over specific time periods. Particularly, in the study of cell division, under normal conditions or after drug treatments, this methodology can provide essential data that is impossible to obtain from conventional cell-fixed-based assays. In the context of evaluating cell fate after antimitotic drug treatment, time-lapse imaging provides valuable insights into the mechanisms of action of these compounds. Antimitotic drugs belong to a class of anticancer compounds that interfere with mitosis by targeting microtubules or specific kinases and molecular motors involved in the mitotic apparatus. This chapter aims to highlight the significant advantage of using time-lapse microscopy to assess the effects of antimitotic drugs on cell behavior and fates. We describe the full protocol for time-lapse imaging, using the human lung adenocarcinoma A549 cell line as a model, after exposure to the antimitotic BI2536, a potent inhibitor of polo-like kinase 1 (PLK1). This description includes software for imaging acquisition and data analysis.

Key words Time-lapse imaging, Cell tracking, Cell dynamics, Cell fate, Antimitotic agents, BI2536, Polo-like kinase 1

1 Introduction

The mitotic machinery is an extensively explored target for studying potential anticancer treatments. The class of drugs that target the mitotic machinery is called antimitotic drugs and can be divided into (i) conventional microtubule-targeting agents (MTAs) and (ii) non-microtubule-targeting agents. MTAs are a well-known class of antimitotic drugs which target microtubules and are currently in use in the clinic. One of the best characterized MTA is paclitaxel (or taxol), widely used in several types of solid tumors including from breast, lung, ovarian, pancreatic, among others [1]. Despite the MTAs clinical history of success, the treatment with these drugs has diverse side effects, namely, toxicity and acquired resistance.

Victor M. Bolanos-Garcia (ed.), *Mitotic Exit: Methods and Protocols*, Methods in Molecular Biology, vol. 2874,
https://doi.org/10.1007/978-1-0716-4236-8_10,

The non-microtubule-targeting agents are a group of small molecules that inhibit specific mitotic proteins with key roles in mitosis, typically kinases and molecular motors. Many of these small molecules reached clinical trials, namely, inhibitors of MPS-1 (monopolar spindle 1), PLK1 (polo-like kinase 1), Aurora kinases, KLP1 (kinesin-like protein 1), and CENP-E (centromeric protein E) [2].

Antimitotic drugs can induce a variety of cell fates. The significant challenge in in vitro studies has been to understand what occurs to the cell, both spatially and temporally, after exposure to the drug, including the elucidation of their mechanisms of action [3]. There are two main potential outcomes for cells that are arrested in mitosis under antimitotic treatment: i) the cells might die in mitosis, an event that can easily be seen by light microscopy since dying cells are seen to contract, bleb, and/or break; or ii) a process known as mitotic slippage allows cells to leave mitosis and transition back to interphase without cytokinesis, an event that is easily recognized by the decondensation of chromosomes and/or the flattening of the mitotic cell, with cells that escape mitosis usually having irregular and multi-lobed nuclei (such cells have an increased probability of accumulating errors that can ultimately lead to their death or be associated with drug resistance) (Fig. 1) [4]. Cell behavior after antimitotic treatment varies depending on the antimitotic drug used, the cell line, and even among cells of the same line [5].

Live cell imaging has been commonly employed to analyze the time-dependent dynamic nature of cell behavior, and various imaging techniques can be applied, including brightfield, fluorescence, and confocal microscopy. Through time-lapse imaging, we can observe and track individual cells in real-time, enabling the visualization of crucial events including cell division, cell motility, and cell death. This technique enhances the comprehensive analysis of cellular responses to antimitotic drugs, providing a detailed understanding of the drug's impact on cell cycle progression and the determination of cell fate [6, 7].

Time-lapse imaging offers a unique opportunity to investigate cellular responses to antimitotic drugs, contributing to a more comprehensive understanding of cancer biology and potential avenues for personalized medicine approaches. The comprehensive protocol presented in this chapter for brightfield time-lapse imaging of A549 cancer cell line, using differential interference contrast (DIC) microscopy, provides a practical guide for researchers interested in tracking cell dynamics, including chromosome movement, and cell fate after prolonged mitotic arrest. We describe a protocol for directly observing and studying the fate of cells that experience extended mitotic arrest in response to the antimitotic drug BI2536, a small-molecule inhibitor that targets PLK1 [8]. PLK1 is a serine/threonine-protein kinase with specific roles during mitotic

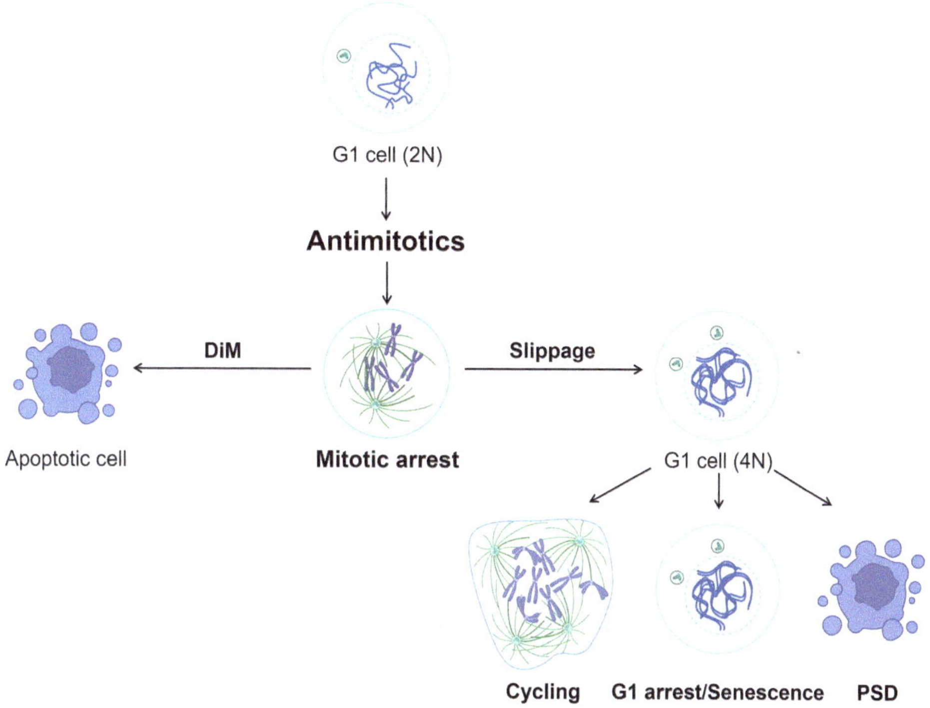

Fig. 1 Cell fate after mitotic arrest induced by antimitotics. A diploid cell (represented as a G1 cell with 2 N DNA content) treated with antimitotics that induce prolonged mitotic arrest may undergo death in mitosis (DiM), or slippage. The resulting slipped tetraploid cell (represented as a G1 cell with 4 N DNA content) may arrest in interphase (G1 arrest/senescence), undergo post-slippage death (PSD), or continue cycling. Created in BioRender.com

progression, including centrosome maturation, spindle assembly checkpoint, chromosome segregation and cytokinesis, and is overexpressed in multiple cancers [9, 10]. Inhibition of PLK1, particularly by BI2536, results in monopolar spindle formation and extended mitotic arrest expected to culminate into cancer cell death [8]. Nevertheless, the fate of cells arrested in mitosis due to BI2536 varies considerably: While a portion of cells undergoes mitotic death, a significant fraction experiences mitotic slippage, with the latter group possibly resulting in cell death or survival [7].

2 Materials

2.1 Cell Seeding

1. Cell line: human lung adenocarcinoma A549.
2. Complete culture medium: RPMI-1640 supplemented with 10% FBS and 1% penicillin-streptomycin.
3. Phosphate-buffered saline (PBS)1x: 1.37 M NaCl, 27 mM KCl, 100 mM Na2HPO$_4$, 18 mM KH_2PO_4, pH 7.4.

4. Trypsin-like protease TrypLE Express.
5. T25 culture flasks.
6. Serologic pipettes: 2, 5, and 10 mL.
7. Pipette controller.
8. Ethanol 70% (v/v) and absolute.
9. Extran 10%.
10. Waste container.
11. Conic tube 15 mL.
12. Centrifuge tube 1.5 mL.
13. Sterile Water.
14. LabTek II chambered cover glass.
15. Trypan Blue Exclusion dye.
16. Neubauer counting chamber.
17. Micropipettes: P1000, P200, P20.
18. Micropipette Tips.
19. Cell incubator.
20. Water-bath.
21. Cell counter.
22. Inverted microscope (in our case: primo Vert, Carl Zeiss).

2.2 Antimitotic Drugs

1. 10 mM BI2536 in DMSO. To avoid loss of compound activity, prepare several aliquots and store at −20 °C.
2. Dimethyl sulfoxide (DMSO).

2.3 Microscopy

1. Axio Observer Z.1 SD Inverted Light Microscope (Carl Zeiss) with an incubation chamber with temperature, humidity, and CO_2 regulation.
2. AxioVision software.
3. Coverslip (24 × 50 mm).
4. Immersion oil.

2.4 Data Analysis

1. ImageJ software (https://imagej.nih.gov/).
2. Statistical software for data import and analysis.

3 Methods

3.1 Cell Seeding

This protocol can be adapted for any adherent non-tumor or tumor cell line. The culture medium and certain details of the cell sub-culturing procedure may vary depending on the specific cell line (*see* **Notes 1** and **2**). It is essential to verify the characteristics and handling instructions for the selected cell line before proceeding with the protocol.

Perform all procedures in a Biosafety Cabinet Level II to maintain a sterile environment and prevent cell contamination (see **Note 3**). Ensure that all solutions are pre-warmed at 37 °C before use.

1. Remove the medium from the culture flask, wash the cells with 1.5 mL of 1x PBS, and add 500 μL of TrypLE Express reagent to promote cell detachment. Place the flask in an incubator at 37 °C with 5% CO_2. Wait for approximately 3–5 min (the time may vary depending on the cell type; confirm complete cell detachment using a phase contrast microscope).
2. Add 3 mL of culture medium to the flask to halt enzymatic activity, and transfer the cell suspension to a 15 mL conical tube. Prepare the Neubauer counting chamber. In an Eppendorf tube, combine 30 μL of Trypan Blue with 30 μL of the cell suspension (1:1 ratio). Withdraw 10 μL of the mixture and count the cells using the Neubauer chamber. Use the following formula to calculate cell density (total cells/mL):

$$\text{Cellular Density (cells/mL)} = \frac{\text{number of viable cells}}{\text{number of counted quadrants}} \times \text{dilution factor} \times 10.000.$$

3. Prepare the dilution to achieve the desired initial cell concentration. For A549 cells, we aimed for a concentration of 0.09×10^6 cells/mL.
4. Clean the LabTek II wells by washing them first with 70% ethanol, followed by 96% ethanol, and then allow them to dry. Add autoclaved water to three of the LabTek wells. The water helps maintain relative humidity and prevent the precipitation of culture medium components. Do not substitute PBS for autoclaved water to avoid evaporation and subsequent salt precipitation. In the remaining wells, add 1 mL of medium with the desired cell concentration and place LabTek in the cell incubator for 24 h.

3.2 Drug Treatment

1. Preform a time-lapse imaging (steps described below) on both control untreated and cells treated with DMSO (typically at a concentration < 1%) to verify that the imaged cells maintain their health, without signs of stress, throughout the entire imaging experiment (*see* **Note 4**).
2. Just before starting the time-lapse imaging, dilute BI2536 at the desired concentration in a total volume of 1 mL of fresh culture medium and then add it to the cells into the labTek.

3.3 Microscope Setup and Time-Lapse Imaging

The image acquisition setup relies on three key components: an inverted microscope equipped with an automated stage, automated fluorescence and brightfield shutters, and automated filter wheels; a cell incubator that fits onto the microscope stage and offers an optimal on-stage cell culture environment by maintaining the temperature, humidity, and pH during time-lapse imaging (*see* **Note 5**); and a software that controls the stage, shutters, and filter wheels.

1. Turn on the microscope, the TempModule S1, the heating unit XL S, and the Heating Insert P Lab-Tek TM compact S1, setting them to 37 °C, and allow the temperature to stabilize (this typically takes approximately 2 hours) (Fig. 2).
2. Perform the steps outlined in point 2 of Subheading 3.2 (Drug treatment).
3. Configure the CO_2 to 5% using the CO_2 Module S1 and CO_2 cover HP heating P insert.
4. Replace the LabTek lid with a coverslip previously cleaned with 70% (v/v) ethanol and place the Labtek into the Insert P.

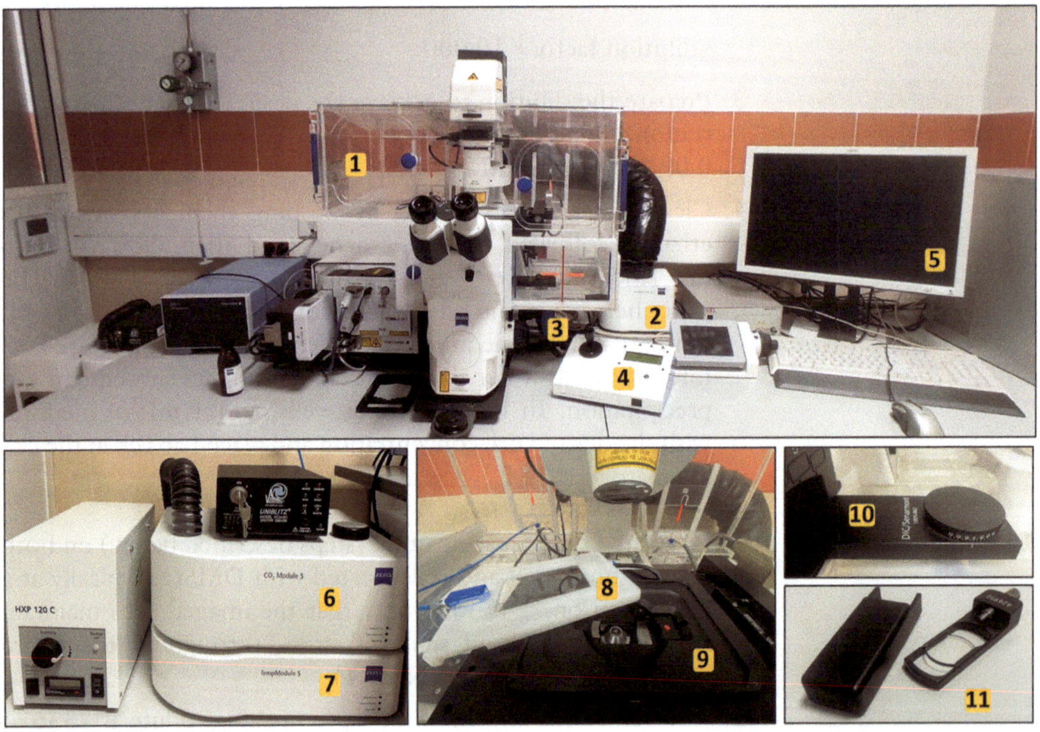

Fig. 2 Axio Observer Z.1 SD Inverted Light Microscope (Carl Zeiss). (1) Incubation chamber; (2) Heating unit XL S; (3) Camera; (4) Motorized x/y stage controller; (5) Computer; (6) CO_2 Module S1; (7) TempModule S1; (8) CO_2 cover HP heating P insert; (9) Heating Insert P Lab-Tek TM compact S1; (10) Polarizer/Analyzer slider; (11) DIC prism

5. Set up the transmitted light and, using the minimum objective magnification, perform the KÖHLER illumination.
6. Switch to the 63x plan objective, add immersion oil, and focus. Depending on the required optical resolution, other objectives may be employed (*see* **Note 6**). Perform the KÖHLER illumination for this objective. It is essential to place the immersion oil into the incubator heating chamber during microscope stabilization to warm it to 37 °C. Using immersion oil at room temperature can adversely affect focus and resolution during imaging (*see* **Note** 7).
7. Remove the objective protection and place the prism.
8. Insert the polarizer and turn on the DIC.
9. Turn off the light while opening the Multidimensional

   Acquisition window in the software and select the appropriate parameters for your experiment (specify the Z-stack positions, exposure time (*see* **Note 8**), and the interval time (*see* **Note 9**) between image collections). Typically, normal mitosis in A549 cells lasts 40–50 min. Therefore, using 5 minute intervals between image acquisitions for the 72 h experiment duration will provide sufficient data on the cell division process (*see* **Notes 10** and **11**).
10. Using the acquisition software, select several non-overlapping fields that are ideal for imaging and subsequent analysis, where cells have adhered well to the LabTek slide (*see* **Note 12**), achieving approximately 70% confluent, and avoiding any clumps.
11. Enable the microscope's autofocusing feature to ensure that all designated points remain in focus during the course of the experiment.
12. Click "start" to initiate acquisition.
13. Periodically verify that the image acquisition is proceeding smoothly over the course of acquisition.

3.4 Cell Fate Tracking and Data Analysis After BI2536 Treatment

1. Save the image file in zvi format or other formats that are manageable.
2. Open the file in the ImageJ program.

   Track the cell cycle progression of individual cells in a field based on morphological changes, as illustrated and described in Fig. 3. Observe the cells from frame to frame. In interphase, cells exhibit a flattened morphology and round-shaped nucleus containing decondensed chromosomes. The onset of mitosis is marked by changes in cell shape, resulting in a more rounded appearance, along with the migration of condensed and well-individualized chromosomes toward the cell's center. Exit from mitosis is characterized by the cells reverting to their flattened morphology and losing its rounded appearance.

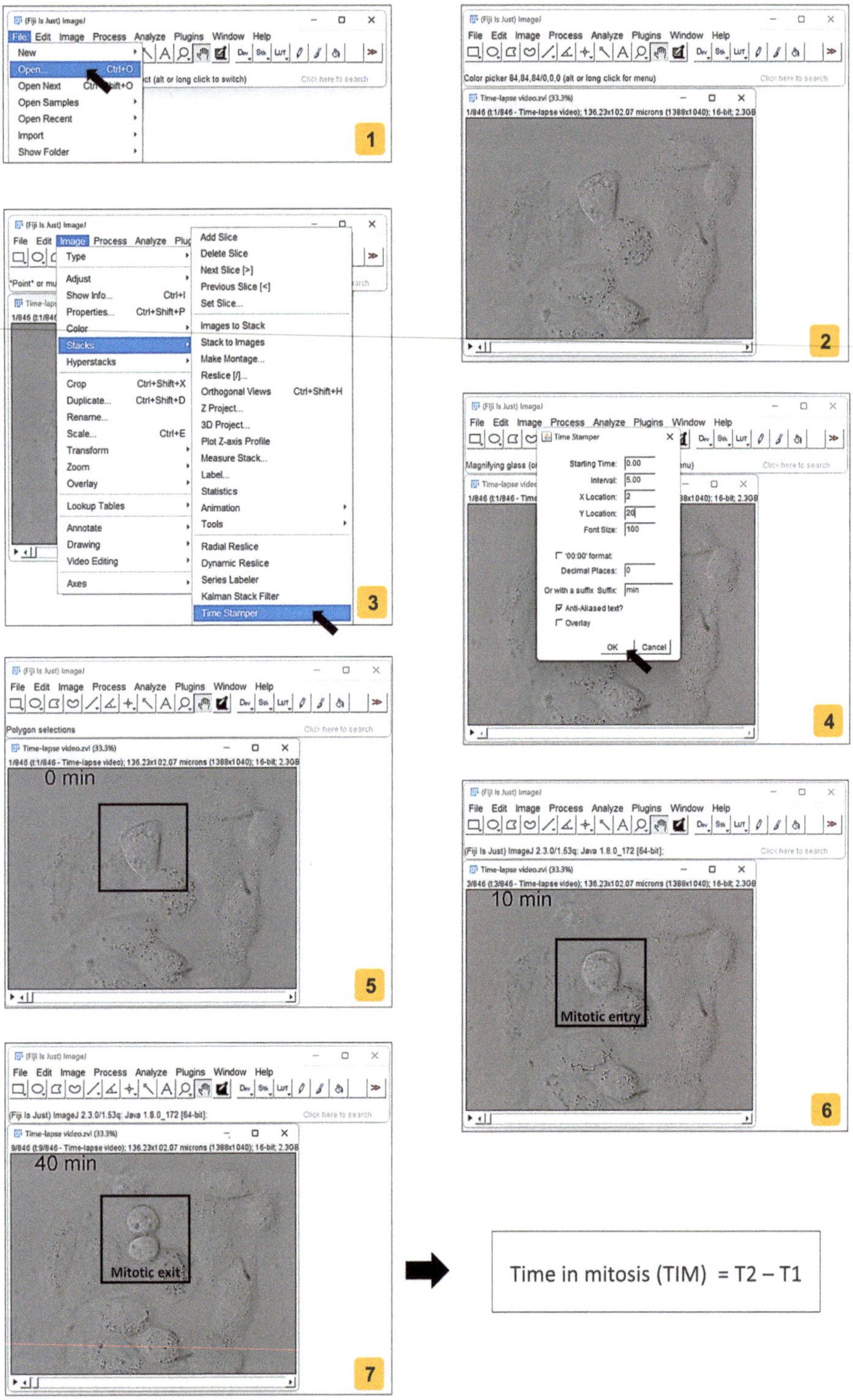

Fig. 3 Sequential steps to analyze time-lapse data using the software ImageJ. (1–2) Open your ImageJ software and import your time-lapse data using: File > Open > Ok; (3) To insert a time stamper, go to: Image > Stacks > Time Stamper and (4): choose the format, interval length, X and Y location and font size, and select Ok; (5–7) Track each cell from the onset of mitosis (time 1) to exit from mitosis (time 2), based on

3. Control cells should align their chromosomes efficiently and divide into two daughters within approximately 1 h (Fig. 4).
4. BI2536-treated cells remain in mitosis with a rounded morphology and condensed chromosomes (Fig. 4) (*see* **Note 13**).
   (a) Mitotically arrested cells undergoing cell death during mitosis (**DiM**) can be identified by exhibiting blebbing, shrinking, and/or fragmenting (Fig. 4).
   (b) Mitotically arrested cells undergoing mitotic slippage can be identified by their reversion to an interphase appearance, featuring decondensed chromosomes, without progressing to cell division. This often results in the formation of large cells (Fig. 4). To determine the fate of cells following mitotic slippage, utilize the aforementioned criteria of cell blebbing, shrinking, or fragmentation for identifying post-slippage death (**PSD,** Fig. 4), as well as the previously mentioned criteria of flattened shape and larger size indicative of the interphase state or G1 arrest/senescence to identify post-slippage survival (**PSS,** Fig. 4). It is worth noting that some of the latter cells may exhibit a re-entry into the cell cycle, as evidenced by their rounded morphology.
5. Continue tracking cells within the current field of view. Repeat the analysis with the other fields of view once all cells within a field are tracked.
6. The data obtained from time-lapse sequences can be represented by different types of graphs (Fig. 4) such us histograms and scatter plot.

4 Notes

1. **Cell line doubling time:** Different cell lines have distinct doubling times, which can range from a few hours to several days. It is crucial to account for these differences when planning and conducting time-lapse imaging experiments. View detailed information about the selected cell line. Typically, cell line manufacturer websites contain relevant information for adequate handling.

Fig. 3 (continued) morphological changes. Time 1 is primarily determined by cell rounding and DNA condensation and time 2 is characterized by physical division into two daughter cells and by cell flattening. The mitosis duration is the time that the cell spends from time 1 to time 2 (TIM = T2—T1). Alternatively, you can assess the duration of mitosis (or the transition between stages) by calculating the time between frames multiplied by the number of frames during which cells are in mitosis or a specific stage. [8] Save the movie as . tiff or as .avi for use in presentations and publications or other format according to your preferences

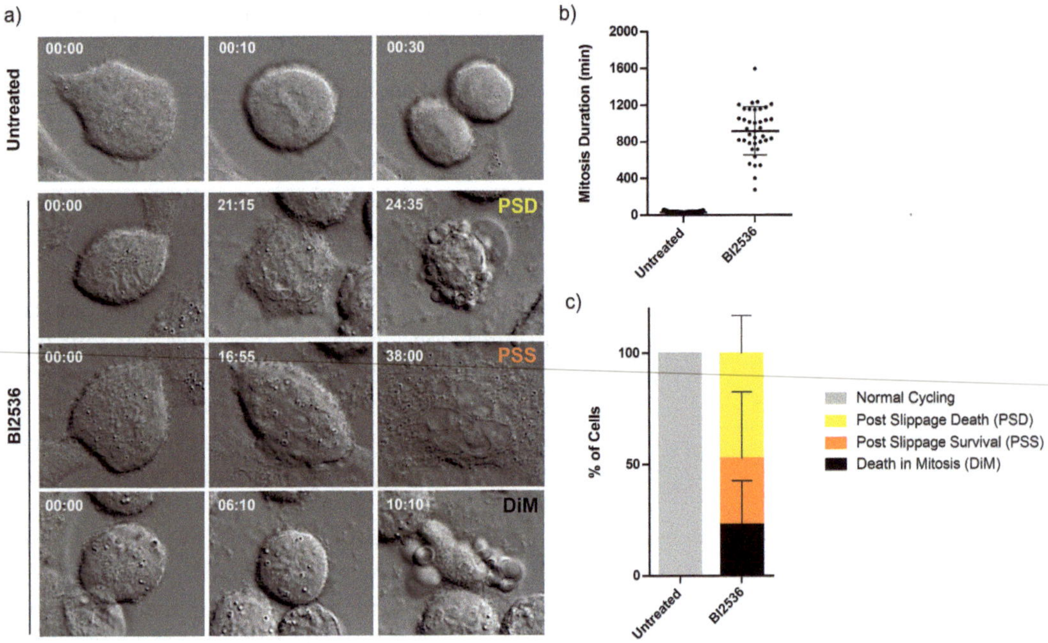

Fig. 4 Cell fates observed after cell treatment with the PLK-1 inhibitor BI2536. **(a)** Representative time-lapse sequences of untreated and BI2536-treated cells. Untreated cells (n = 27) undertake mitosis for about 30 min. In contrast, BI2536-treated cells (n = 38) arrest in mitosis during several hours (918.2 ± 263.5 min) and die in mitosis (DiM) or undergo slippage followed by death (PSD) or remain alive (PSS). Numbers indicate time in 00 h:00 min. **(b)** Quantification of mitosis duration in untreated and BI2536-treated cells, as determined by time-lapse microscopy. Each dot represents one cell. **(c)** Quantification of cell fate in untreated and BI2536-treated cells. Percentage of cells undergoing post-slippage death (PSD), cells that remain alive after slippage (post-slippage survival, PSS), cells that die in mitosis (DiM), and cells with normal cycling are shown. The error bars represent mean ± SD of three independent experiments

2. **Appropriate culture conditions:** Maintain optimal culture conditions for cells, including suitable media, temperature, and relative humidity. Ensure that the cells are placed in a controlled environment to prevent fluctuations that may affect their behavior during imaging. Again, view detailed information about the selected cell line. Typically, cell lines manufacturer websites contain relevant information for adequate handling.
3. **Cell viability:** Assess cells viability before initiating time-lapse imaging. Ensure that the cells are healthy, exhibit appropriate morphology, and are not compromised by factors such as contamination or over-confluence. Cell density has profound effect on the growth of the cell line and, in consequence, on the fidelity of the experimental results.
4. **Experimental controls:** Include appropriate controls in your experiments, such as untreated cells or cells treated with vehicle only, to distinguish drug-specific effects from other factors that may influence cell behavior during imaging.

5. **Environmental factors:** Minimize external disturbances such as vibrations, temperature fluctuations, or airflow that can affect the stability of the imaging setup and introduce artifacts into the data.
6. **A high objective** (such as 40 or 63x) should be used to create a time-lapse figure in order to more effectively depict the mitotic figs. A major limitation is that only a few cells are analyzed. To increase the number of cells that can be analyzed, the entire well can be imaged; alternatively, the lowest objective (20x) that nevertheless permits unmistakable identification of mitotic formations should be employed. A second significant limitation of this approach is how time- and labor-intensive it is to manually track cell fates. However, there are cell-tracking algorithms that could be enhanced for automated image interpretation.
7. **Focus stability:** A stable focus should be maintained throughout the imaging acquisition period. Unstable focus can lead to blurred images, hindering accurate analysis of cellular events.
8. **Phototoxicity:** Minimize the impact of phototoxicity on cells during time-lapse imaging. Prolonged exposure to intense light can affect cell behavior and viability. Consider using low-intensity illumination or limiting the duration of imaging sessions.
9. **Time interval selection:** Determine the appropriate time intervals for capturing images based on the dynamics of the cellular process of interest. Balancing shorter intervals for capturing fine details and longer intervals to minimize phototoxicity is essential.
10. **Image acquisition settings:** Optimize imaging parameters, such as exposure time and image resolution, to obtain clear and high-quality images. Experiment with different settings to achieve the best balance between image quality and cellular viability.
11. **The risk of photodamage** to the cells, which prevents division, can rise with short time intervals (< 2 min). Data loss may occur over lengthy periods of time (> 10 min). Anaphase, slippage, or apoptosis are relatively quick events, and would not be seen at intervals of 15 min. In our hands and microscope setting, 5 min intervals are a reasonable time compromise for dividing cells, however this will need to be adjusted based on the sample type and temperature. Also, one should avoid taking a full Z stack to minimize photodamage, one approach is to take a single image or a small Z stack at each time point.
12. **Sample movement:** Prevent sample drift or movement during imaging, as it can disrupt the analysis. Ensure that the cells are immobilized appropriately using techniques such as specialized culture dishes, micropatterned substrates, or chemical fixation.

13. It is crucial to consider that certain antimitotic drugs, especially those that disable the spindle assembly checkpoint, lead to **premature exit from mitosis** rather than mitotic arrest. As a result, the duration of mitosis in treated cells is significantly shortened. Therefore, the protocol for tracking the fates of mitotic cells outlined in this chapter can still be used, but it requires adjustments to the interval time between image collections.

Acknowledgments

This work was supported by CESPU—Cooperativa de Ensino Superior Politécnico e Universitário Crl (Grants Ref. SGA4Cancer-GI2-CESPU-2022, and Ref. upPTXovcar-GI2-CESPU-2022) to H.B. and P.M.A.S. B.P. gratefully acknowledges CESPU (BD/CBAS/CESPU/01/2020) and Fundação para a Ciência e a Tecnologia (FCT) (2022.09451.BD) for financial support. J.P.N.S. gratefully acknowledges CESPU (BD/CBAS/CESPU/01/2021) for financial support.

References

1. Van Vuuren RJ, Visagie MH, Theron AE et al (2015) Antimitotic drugs in the treatment of cancer. Cancer Chemother Pharmacol 76:1101–1112
2. Novais P, Silva MA, Amorim I et al (2021) Second-generation Antimitotics in cancer clinical trials. Pharmaceutics 13:1011
3. Piltti KM, Cummings BJ, Carta K et al (2018) Live-cell time-lapse imaging and single-cell tracking of in vitro cultured neural stem cells – Tools for analyzing dynamics of cell cycle, migration, and lineage selection. Methods 133:81–90
4. Henriques AC, Ribeiro D, Pedrosa J et al (2019) Mitosis inhibitors in anticancer therapy: when blocking the exit becomes a solution. Cancer Lett 440–441:64–81
5. Gascoigne KE, Taylor SS (2008) Cancer cells display profound intra- and interline variation following prolonged exposure to antimitotic drugs. Cancer Cell 14:111–122
6. Mackay DR, Ullman KS, Rodesch CK (2010) Time-lapse imaging of mitosis after siRNA transfection. J Vis Exp 8–10
7. Pinto B, Novais P, Henriques AC et al (2022) Navitoclax enhances the therapeutic effects of PLK1 targeting on lung cancer cells in 2D and 3D culture systems. Pharmaceutics 14:1–17
8. Steegmaier M, Hoffmann M, Baum A et al (2007) BI 2536, a potent and selective inhibitor of polo-like kinase 1, inhibits tumor growth in vivo. Curr Biol 17:316–322
9. Colicino EG, Hehnly H (2018) Regulating a key mitotic regulator, polo-like kinase 1 (PLK1). Cytoskeleton (Hoboken) 75:481–494
10. Liu Z, Sun Q, Wang X (2017) PLK1, a potential target for cancer therapy. Transl Oncol 10:22–32

Chapter 11

The Role of Polo-Like Kinase 1 (PLK1) O-GlcNAcylation in Mitosis

Jie Li, Guangcan Shao, Bin Peng, Xingzhi Xu, Meng-Qiu Dong, and Jing Li

Abstract

Polo-like kinase 1 (PLK1) is a crucial mitotic kinase that is implicated in various aspects of cell cycle. Many post-translational modifications have been identified on PLK1 to regulate its activation, stability, and localization. PLK1 has been shown previously to colocalize with the O-linked β-N-acetylglucosamine (O-GlcNAc) transferase (OGT), and OGT regulates PLK1 stability. In our recent work, we show that PLK1 is O-GlcNAcylated by click chemistry. Using stepped collisional energy/higher energy collision dissociation mass spectrometry, we mapped the PLK1 O-GlcNAc site to be T291. We further utilized fluorescent activated cell sorting and time-lapse microscopy to assess the mitotic defects of PLK1 O-GlcNAc mutants. In vivo studies in mouse xenograft demonstrated that it promoted uterine cancer tumorigenesis. In this chapter, we delineate the methodologies we used in studying PLK1 O-GlcNAcylation, including click chemistry, stepped collisional energy/higher energy collision dissociation mass spectrometry, fluorescent activated cell sorting, time-lapse microscopy, and mouse xenograft assays.

Key words O-GlcNAcylation, PLK1, Mitosis, Mass spectrometry, Uterine carcinoma

1 Introduction

O-linked β-N-acetylglucosamine (O-GlcNAc) modification is a quintessential glycosylation that decorates the Ser/Thr residues of intracellular proteins. It occurs in the cytosol, nucleus, and mitochondria [1]. Functionally, it crosstalks with other post-translational modifications, such as phosphorylation and ubiquitination, to regulate protein localization, stability, and protein-protein interactions [2]. A role of O-GlcNAc in the cell cycle regulation has been long appreciated, as deletion of the O-GlcNAc transferase (OGT) impedes cell cycle progression [3]. Moreover, many efforts have been devoted to pinpoint the mitotic substrates of OGT as well as how OGT is regulated by the cell division machinery [4, 5].

Victor M. Bolanos-Garcia (ed.), *Mitotic Exit: Methods and Protocols*, Methods in Molecular Biology, vol. 2874,
https://doi.org/10.1007/978-1-0716-4236-8_11,

Although important for many biological signaling pathways, the identification of OGT substrates and their exact O-GlcNAcylation sites are hindered by its intrinsic nature: low stoichiometry, high lability, and being easily lost during tradition mass spectrometry (MS) analysis. Recent advancement in MS, especially higher energy collision dissociation (HCD) and electron transfer dissociation (ETD) MS, has witnessed great strides in O-GlcNAc glycomics [6, 7], as well as single protein O-GlcNAcylation analysis.

Polo-like kinase 1 (PLK1) is a mitotic master kinase that orchestrates various aspects of cell cycle: replication, mitotic entry, centrosome separation, sister chromatin segregation and cytokinesis [8]. Due to its pivotal role, its inhibitors are tested in clinical settings [9]. In a previous investigation, PLK1 has been demonstrated to colocalize with OGT and OGT regulates PLK1 stability [4]. In our recent work [10], we demonstrate what PLK1 is O-GlcNAcylated. Using stepped collisional energy (sce) HCD MS, we mapped the glycosite to be T291. O-GlcNAc regulates PLK1 stability and thus cell cycle progression, as demonstrated by fluorescent activated cell sorting (FACS) and time-lapse microscopy. As PLK1-T291N is also a uterine carcinoma mutant in the Cancer Genome Atlas, we further performed mouse xenograft assays and found that PLK1-T291A and -T291N promote uterine cancer tumorigenesis [10]. Here we delineate the methods we used in our research, including click chemistry, sce-HCD MS, FACS, time-lapse microscopy, and mouse xenograft assays.

2 Materials

2.1 Click Chemistry

1. 293T cells were purchased from ATCC.
2. 200 mM $Ac_36AzGlcNAc$ (Delta).
3. 20 mM Thiamet-G (Sigma-Aldrich).
4. 1 × PBS.
5. 150 mM Lysis Buffer: 150 mM NaCl, 1 M Tris–HCl (pH 7.5), 0.5 M EDTA, 10% NP-40, and 1 × protease inhibitor cocktail.
6. BCA Protein Assay Kit (Sigma-Aldrich).
7. Methanol.
8. Chloroform.
9. Labeling Buffer: 8 mM Urea, 10 mM HEPES (pH 7.9), 1 × Halt™ Protease and Phosphatase Inhibitor Cocktail, and 1 mM PMSF.
10. 50 mM Biotin-PEG4-DBCO (Du You You).
11. Protein Resoluble Buffer: 1% SDS, 50 mM Na_2HPO_4 (pH 8.0), and 100 mM NaCl.

12. Neutralizing Buffer: 6% NP-40, 50 mM Na_2HPO_4 (pH 8.0), and 100 mM NaCl.
13. Streptavidin-Beads (Sigma-Aldrich).
14. Low-Salt Buffer: 50 mM Na_2HPO_4 (pH 8.0), 100 mM NaCl, 0.1% SDS, 1% Triton X-100, and 0.5% sodium deoxycholate.
15. High-Salt Buffer: 50 mM Na_2HPO_4 (pH 8.0), 500 mM NaCl, and 1% Triton X-100.
16. 2 × Sample Buffer: 100 mM Tris–HCl (pH 6.8), 20% Glycerol, 4% SDS, 2% 2-Mercaptoethanol, and 0.2% Bromophenol Blue.

2.2 sce-HCD MS

1. 10% SDS-PAGE.
2. Coomassie Brilliant Blue Staining Buffer: 1 g Coomassie Brilliant Blue R-250, 450 mL methanol, 100 mL glacial acetic acid, 450 mL ddH_2O.
3. Coomassie Brilliant Blue Destaining Buffer/glacial acetic acid/methanol/ddH_2O = 1/1/1/8.
4. 100 mM NH_4HCO_3.
5. Reduction Reaction Buffer: contains 20 mM NH_4HCO_3 and 5 mM TCEP.
6. Alkylation Reaction Buffer: contains 10 mM iodoacetamide and 20 mM NH_4HCO_3.
7. Water (LC-Mass Spectrometry Grade).
8. Acetonitrile (LC-Mass Spectrometry Grade).
9. 50 mM NH_4HCO_3.
10. Trypsin (LC-Mass Spectrometry Grade) (0.5 μg/μL).
11. 50% Acetonitrile (LC-Mass Spectrometry Grade).
12. 75% Acetonitrile (LC-Mass Spectrometry Grade).
13. 5% TFA (LC-Mass Spectrometry Grade).
14. 1% TFA (LC-Mass Spectrometry Grade).
15. Easy-nLC 1000 II HPLC (Thermo Fisher Scientific).
16. Q-Exactive HF mass spectrometer (Thermo Fisher Scientific).
17. Pre-column (75 μm ID, 6 cm long, packed with ODS-AQ 10 μm, 120 Å beads from YMC Co., Ltd.).
18. Analytical column (75 μm ID, 12 cm long, packed with Luna C18 1.9 μm 100 Å resin from Welch Materials).
19. Buffer A: 0.1% formic acid in ddH_2O.
20. Buffer B: 0.1% formic acid in acetonitrile.
21. pFind 3 software.

2.3 FACS

1. HeLa cells stably expressing GFP-H2B were infected with lentiviral plasmids FLAG-VEC, PLK1-WT, -T291A, or -T291N constructs.
2. shRNA: sh*PLK1* UTR CCGGAGCTGC ATCATCCTTG CAGGTCTCGA GACCT GCAAG GATGATGCAG CTTTTTT.
3. 0.25% trypsin.
4. Cold 1 × PBS.
5. 0.25% Triton X-100.
6. Cold ethanol.
7. Blocking Buffer: 2% BSA.
8. Propidium Iodide (PI) staining buffer (Beyotime).
9. 10 mg/mL RNase A (Beyotime).
10. 300 mesh size nylon filter membrane.
11. CytoFLEX S, Beckman Coulter.
12. FlowJo Analysis Software.

2.4 Time-Lapse Microscopy

1. HeLa cells stably expressing GFP-H2B were infected with lentiviral plasmids FLAG-VEC, PLK1-WT, -T291A or -T291N constructs.
2. shRNA: sh*PLK1* UTR CCGGAGCTGC ATCATCCTTG CAGGTCTCGA GACCT GCAAG GATGATGCAG CTTTTTT.
3. Okola Stage Top Incubator (Oko-Lab).
4. Andor Dragonfly confocal microscope (Oxford Instruments).
5. Imaris Microscopy Image Analysis Software (Oxford Instruments).

2.5 Mouse Xenograft

1. HeLa cells infected with lentiviral plasmids FLAG-VEC, PLK1-WT, -T291A, or -T291N constructs.
2. Matrigel matrix (Corning).
3. 0.25% Trypsin.
4. 1 × PBS.
5. 1 mL syringe.
6. Nude mice.

3 Methods

3.1 Click Chemistry

1. Transfect the target protein and OGT plasmids into 293T cells and culture in medium at 37 °C.
2. Add $Ac_36AzGlcNAc$ and keep its final concentration at 200 μM 24 h before cell collection.

3. Add Thiamet-G at 10 μM 12 h before cell collection.
4. Collect 5 × 10^6 cells and centrifuge at 500× *g* for 5 min.
5. Resuspend cells in 10 mL 1 × PBS and centrifuge at 500× *g* for 5 min.
6. Resuspend cells in 1.5 mL 150 mM Lysis Buffer and incubate on ice for 1 h.
7. Transfer the lysates to a new 1.5 mL tube and centrifuge at 12,500 rpm at 4 °C.
8. Collect 800 μL supernatant to a new tube for the next step and 20 uL to detect input proteins.
9. Measure the protein concentration by BCA and adjust the sample concentration to 1 mg/mL.
10. Transfer 200 μL solution to a new tube, add 600 μL methanol, 200 μL chloroform, and 450 μL ddH_2O to the tube sequentially and mix it completely.
11. Centrifuge the sample at 22,000× *g* for 15 min, remove the supernatant.
12. Add 450 μL methanol to the tube and mix it completely, centrifuge the sample at 22,000× *g* for 15 min.
13. Remove the supernatant and dry in a vacuum concentrator at 30 °C for 5 min.
14. Re-dissolve precipitated proteins in Labeling Buffer at a concentration of 4 mg/mL.
15. Add Biotin-PEG4-DBCO to the protein solution and keep its final concentration at 50 μM.
16. Protect the reaction from light and place it on a rotary mixer at 37 °C for 4 h.
17. Precipitate the protein again following step 10 to 13.
18. Re-dissolve 500 μg precipitated proteins in 160 μL Protein Resoluble Buffer and add 340 μL Neutralizing Buffer to make the protein at a concentration of 1 mg/mL.
19. Add 30 μL of 50% Strepavidin-Beads suspension to the protein solution and place on a rotary mixer overnight at 4 °C.
20. Centrifuge the sample at 22,000× *g* at 4 °C for 15 min and remove the supernatant.
21. Wash the beads 5 times with Low-Salt Buffer, each with 1 mL.
22. Wash the beads 5 times with High-Salt Buffer, each with 1 mL.
23. Add 2 × Loading Buffer containing 20 mM biotin and denature at 95 °C for 10 min.
24. Detect the amount of target protein by Western Blotting (Fig. 1).

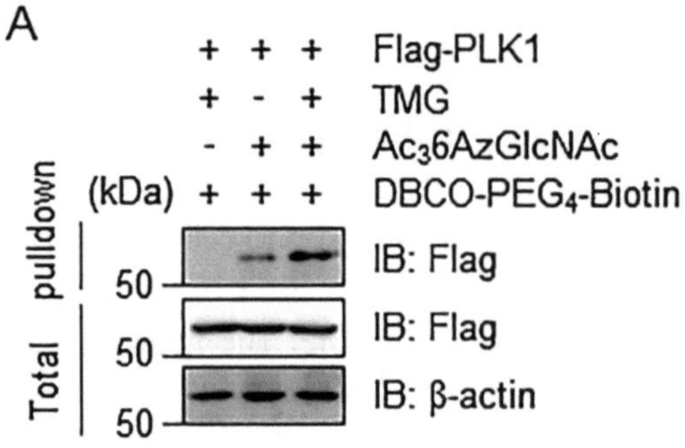

Fig. 1 Click chemistry showed that PLK1 is O-GlcNAcylated. Cells were transfected with Flag-PLK1, then treated with 200 μmol/L $Ac_36AzGlcNAc$ or not treated and 5 μmol/L TMG or not treated as previously described [10]. Click chemistry experiments were carried out

3.2 sce-HCD MS

1. Separate target proteins by SDS-PAGE.
2. Keep the gel in Coomassie Brilliant Blue Staining Buffer and place on a room temperature shaker for 45 min.
3. Remove the Coomassie Brilliant Blue Staining Buffer, keep the gel in Coomassie Brilliant Blue Destaining Buffer, and place on a shaker overnight at room temperature.
4. Cut the SDS-PAGE gel containing the target protein, chop it into 1 mm^3 pieces, and transfer them to a new tube.
5. Wash the gel slices for 20 min in 500 μL 100 mM NH_4HCO_3, repeat 2 times.
6. Reduce the samples with 150 μL Reduction Reaction Buffer, incubate at 56 °C for 40 min and discard the supernatant.
7. Add 150 μL Alkylation Reaction Buffer, incubate in dark at room temperature for 1 h and discard the supernatant.
8. Wash gel slices for 20 min with water (LC-Mass Spectrometry Grade), repeat 2 times.
9. Wash the gel slices with 500 μL acetonitrile (LC-Mass Spectrometry Grade) for 1 min and discard the supernatant, repeat 2 times.
10. Dry under reduced pressure for 20 min.
11. Add 50 μL 50 mM NH_4HCO_3 containing 10 ng/μL trypsin (LC-Mass Spectrometry Grade), incubate for 1 h at 37 °C.
12. Add 50 μL 50 mM NH_4HCO_3 to submerge the samples, incubate overnight at 37 °C.
13. Add 100 μL 50% acetonitrile in water containing 5% TFA, treat gel slices on shaker (600 rpm) for 30 min, collect the supernatant.

14. Add 100 μL 75% acetonitrile in water containing 1% TFA, treat gel slices on shaker (600 rpm) for 30 min, collect the supernatant.
15. Combine all extracts, concentrate and dry the extracts under reduced pressure.
16. Reconstitute the samples with 15 μL water (LC-Mass Spectrometry Grade) containing 0.1% TFA.
17. Perform the LC-MS/MS analysis on an Easy-nLC 1000 II HPLC (Thermo Fisher Scientific) coupled to a Q-Exactive HF mass spectrometer (Thermo Fisher Scientific). Load Peptides on a pre-column (75 μm ID, 6 cm long, packed with ODS-AQ 10 μm, 120 Å beads from YMC Co., Ltd.) and further separate on an analytical column (75 μm ID, 12 cm long, packed with Luna C18 1.9 μm 100 Å resin from Welch Materials) using a linear gradient from 100% Buffer A (0.1% formic acid in H_2O) to 30% Buffer B (0.1% formic acid in acetonitrile), 70% buffer A in 75 min at a flow rate of 200 nL/min.
18. Isolate the top 20 most intense precursor ions from each full scan (resolution 120,000) for HCD MS2 (resolution 15,000; three-step normalized collision energy (25, 27, 30)) with a dynamic exclusion time of 60 s. Exclude precursors with a charge state of 1+, 7+ or above, or unassigned.
19. Use the software pFind 3 [11, 12] to identify O-GlcNAcylated peptides by setting a variable modification of 203.0793 Da at S, T. Set the mass accuracy of precursor ions and that of fragment ions at 20 ppm. Filter the results by applying a 1% FDR cutoff at the peptide level and a minimum of one spectrum per peptide. Annotate the MS2 spectra using pLabel [13] (Fig. 2).

3.3 FACS

1. Transfect Flag-VEC, Flag-PLK1-WT, Flag-PLK1-T291A, and Flag-PLK1-T291N plasmids into HeLa cells stably expressing GFP-H2B, followed by knocking down endogenous PLK1 by lentiviral-based shRNA.
2. Collect 5×10^6 cells with 0.25% trypsin in a 1.5 mL tube.
3. Add 1 mL cold 1 × PBS to the tube, centrifuge at 500× *g* for 5 min and discard the top of aqueous phase.
4. Repeat the above wash step again.
5. Add 300 ml cold 1 × PBS to resuspend cells, then add cold ethanol 300 ml, repeat twice until the ethanol concentration increased to 75%, then fix on ice overnight.
6. Centrifuge at 4 °C at 1000× *g* for 2 min and discard the top of aqueous phase.
7. Add 1 mL Blocking Buffer and place on a rotary rocker at 4 °C for 1 h.

A

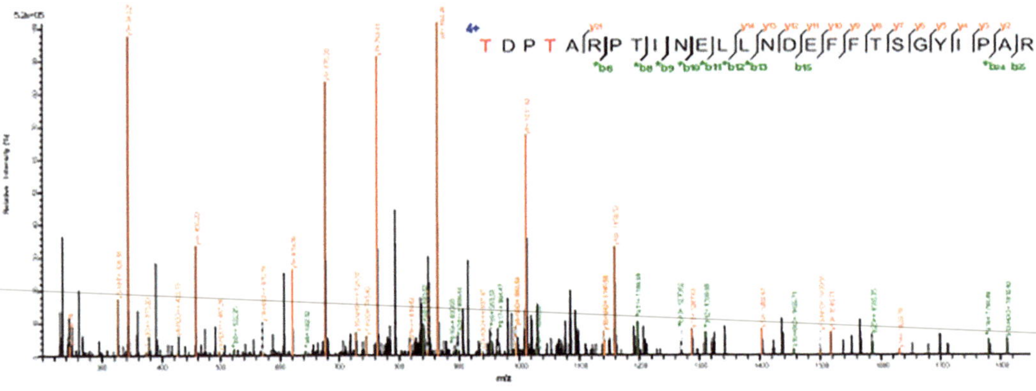

Fig. 2 sce-HCD MS identified that PLK1 is O-GlcNAcylated at Thr-291. Cells were transfected with Flag-PLK1 and enriched for O-GlcNAcylation. Anti-Flag immunoprecipitates were subject to stepped collisional energy/higher energy collisional dissociation (sce-HCD) mass spectrometry analysis. Results showed that Thr291 could be an O-GlcNAcylation site. (* indicates ions with neutral loss of OGT)

8. Centrifuge at 4 °C at 1000× *g* for 2 min and discard the top of the aqueous phase.
9. Add 500 μL Propidium Iodide (PI) staining buffer (final concentration 500 μg/mL) and RNase (final concentration 100 μg/mL) and incubate at room temperature (protected from light) for 30 min.
10. Treat the samples with 300 mesh size nylon filter membrane.
11. Analysis DNA content by using a CytoFLEX flow cytometer (Beckman Coulter) and the cell cycle distribution by using FlowJo software.

3.4 Time-Lapse Microscopy

1. Transfect Flag-VEC, Flag-PLK1-WT, Flag-PLK1-T291A, and Flag-PLK1-T291 plasmids into HeLa cells stably expressing GFP-H2B.
2. Knock down endogenous PLK1 by shRNA targeting the 3′ UTR of PLK1.
3. Place the culture dish in a microincubator to maintain proper environmental conditions (37 °C, high relative humidity (90%), 5% CO_2) using Okola Stage Top Incubator.
4. Acquire the images every 5 min using an Andor Dragonfly confocal microscope.
5. Quantify the durations of prometaphase to anaphase onset in each group using Imaris Microscopy Image Analysis Software, more than 20 cells per group were counted.
6. Identify the misaligned or lagging chromosomes of each cell, more than 20 cells per group were quantified (Fig. 3).

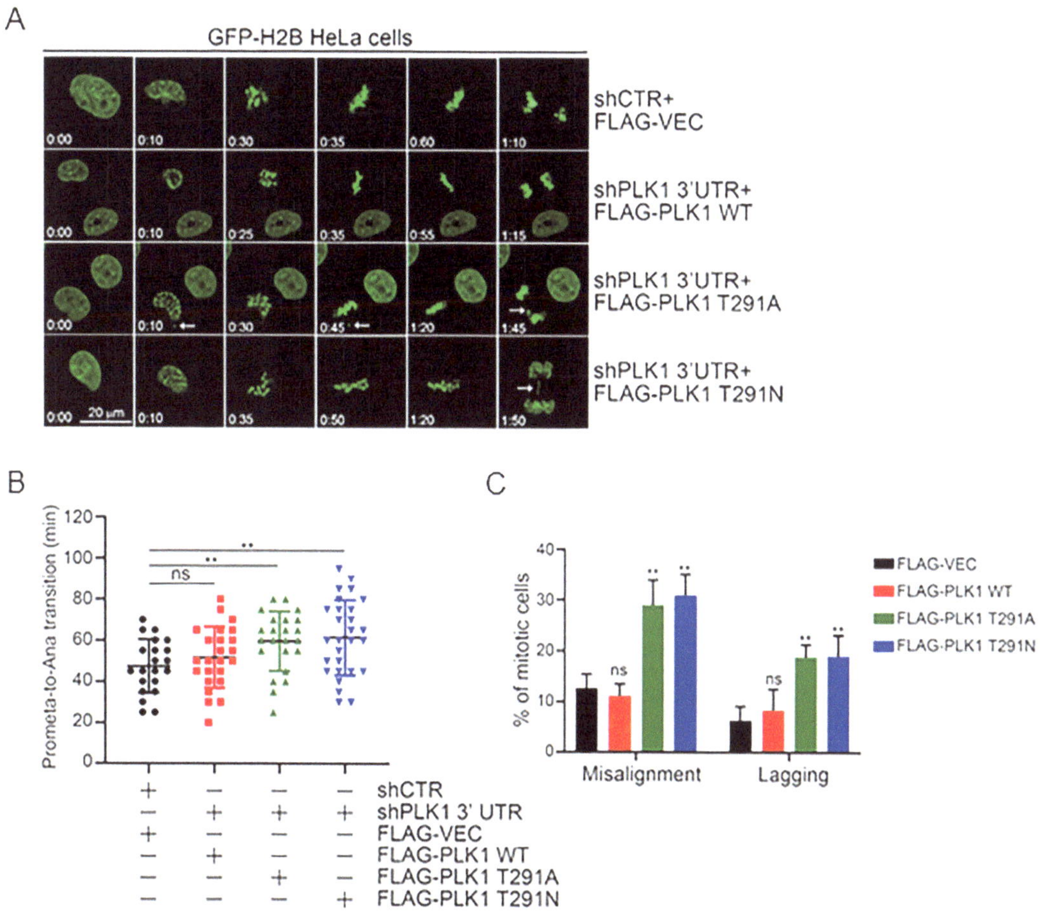

Fig. 3 PLK1 O-GlcNAcylation promotes mitotic progression. GFP-H2B HeLa cells stably expressing FLAG-vec, PLK1-WT, -T291A, or -T291N plasmids were constructed, where endogenous PLK1 was downregulated by shRNA targeting the 3′ UTR of *PLK1*. The cells were subject to time-lapse imaging. Representative time-lapse images are shown with the acquisition time relative to the onset of mitosis

3.5 Mouse Xenograft

1. Establish the cell lines stably expressing Flag-PLK1-WT, Flag-PLK1-T291A, and Flag-PLK1-T291N.
2. Melt the Matrigel at 4 °C in advance before the day to do the inoculation.
3. Collect the cells with 0.25% trypsin and then transfer to a 15 mL tube, centrifuge at 1000× *g* for 5 min.
4. Discard the supernatant, add 1 × PBS to resuspend the cells, and centrifuge at 1000× *g* for 5 min.
5. Discard the supernatant, and add 1 × PBS and Matrigel at a ratio of 1:1 to prepare cell suspension with a cell concentration of 2 × 10^7/mL, which was divided into 1.5 mL EP tubes and store on ice.

6. Use a 1 mL syringe and inoculate 100 uL cell suspension into the right abdominal subcutaneously. Each nude mouse was inoculated with 2×10^6 cells.
7. Measure the tumor volume of mice every 3 days 1 week after injection (tumor volume = (length * width * width) /2).
8. Measure the weight of mice every 3 days 1 week after injection.

4 Notes

To ensure the detection of O-GlcNAcylation signals, several steps need special attention.

1. The efficiency of click chemistry is directly related to the O-GlcNAc abundance of the target protein. When the experimental effect is not in line with expectations, the O-GlcNAc enrichment methods, such as adding Thiamet-G, can be used to improve the results.
2. We actually first used ETD and HCD MS to map PLK1 O-GlcNAcylation sites, but it did not identify any sites in our hands. Usually these two are the first-line methods for O-GlcNAc identification.
3. We used the pFind3 software for O-GlcNAc modified peptides. In this software, there is a parameter where we can set enzyme as trypsin with "Semi-Specific" under "Open Search" when analyzing data. And this is how this semi-specific peptide was identified (Semi-Specific: one terminal of peptide is specific, while the other terminal is not specific). Our follow-up cytology and biochemical assays indicated the importance of Thr291 to O-GlcNAc as well as tumorigenesis.
4. Our sce-HCD identified a peptide modified by 2 O-GlcNAc residues and fragmentation suggested that Thr288 and Thr291 were modified. Our study showed that mutations at Thr291 significantly reduced glycosylation of PLK1, suggesting that this site was modified and that mutation of this region impacted the ability of OGT to glycosylate other residues in PLK1.

Acknowledgement

Jing L. is supported by the National Natural Science Foundation of China (NSFC) fund (32271285 and 31872720) and R & D Program of Beijing Municipal Education Commission (KZ202210028043).

References

1. Yang X, Qian K (2017) Protein O-GlcNAcylation: emerging mechanisms and functions. Nat Rev Mol Cell Biol 18:452–465
2. Hart GW, Slawson C, Ramirez-Correa G, Lagerlof O (2011) Cross talk between O-GlcNAcylation and phosphorylation: roles in signaling, transcription, and chronic disease. Annu Rev Biochem 80:825–858
3. Levine ZG, Walker S (2016) The biochemistry of O-GlcNAc transferase: which functions make it essential in mammalian cells? Annu Rev Biochem 85:631–657
4. Wang Z, Udeshi ND, Slawson C, Compton PD, Sakabe K, Cheung WD, Shabanowitz J, Hunt DF, Hart GW (2010) Extensive crosstalk between O-GlcNAcylation and phosphorylation regulates cytokinesis. Sci Signal 3:ra2
5. Liu C, Li J (2018) O-GlcNAc: a sweetheart of the cell cycle and DNA damage response. Front Endocrinol (Lausanne) 9:415
6. Ma J, Hart GW (2016) Mass spectrometry-based quantitative O-GlcNAcomic analysis. Methods Mol Biol 1410:91–103
7. Ma J, Hart GW (2017) Analysis of protein O-GlcNAcylation by mass spectrometry. Curr Protoc Protein Sci 87:24 10 21–24 10 16
8. Zitouni S, Nabais C, Jana SC, Guerrero A, Bettencourt-Dias M (2014) Polo-like kinases: structural variations lead to multiple functions. Nat Rev Mol Cell Biol 15:433–452
9. Gutteridge RE, Ndiaye MA, Liu X, Ahmad N (2016) Plk1 inhibitors in cancer therapy: from laboratory to clinics. Mol Cancer Ther 15: 1427–1435
10. Yan S, Peng B, Kan S, Shao G, Xiahou Z, Tang X, Chen YX, Dong MQ, Liu X, Xu X, Li J (2023) Polo-like kinase 1 (PLK1) O-GlcNAcylation is essential for dividing mammalian cells and inhibits uterine carcinoma. J Biol Chem 299:102887
11. Chi H, Liu C, Yang H et al (2018) Comprehensive identification of peptides in tandem mass spectra using an efficient open search engine[J]. Nat Biotechnol 36(11):1059–1061
12. Guangcan S, Yong C, Zhenlin C, Chao L, Shangtong L, Hao C, Meng-Qiu D (2021) How to use open-pFind in deep proteomics data analysis?—a protocol for rigorous identification and quantitation of peptides and proteins from mass spectrometry data. Biophys Rep 7(3):207–226
13. Yang B, Wu YJ, Zhu M et al (2012) Identification of cross-linked peptides from complex samples[J]. Nat Methods 9(9):904–906

Chapter 12

Luciferase-Based Reporter Assay for the Assessment of Aurora A-Kinase Activity in Mitotic Cycle

Homa Fatma and Hifzur R. Siddique

Abstract

Luciferase-based reporter assay is an important tool that employs bioluminescence to quickly and precisely investigate the gene of interest's promoter activity by reporter gene expression at the transcriptional level. The promoter of the gene of interest is fused with the reporter gene (a gene that produces luciferase enzymes) and then transfected into the cells. Luciferase is an enzyme that catalyzes a chemical reaction to produce light. The bioluminescence activity of the luciferase gene in the transfected cells is directly proportional to the expression of the gene of interest, which is measured by using a luminometer. In this chapter, we outline the use of a dual-reporter luciferase assay to measure Aurora A kinase activity during the mitotic cycle.

Key words Dual-reporter Luciferase Assay, Mitotic cycle, Aurora A Kinase, Bioluminescence, Promoter, Luminometer

1 Introduction

Dual-luciferase assay has been an important technique to rapidly and accurately determine a promoter activity of a gene of interest by reporter gene expression. The system utilizes bioluminescent firefly luciferase and *Renilla* luciferase to quantify the level of proteins. The bioluminescence generated by the interaction of the bioluminescent reporter with its substrate can be quantified accurately and the bioluminescence activity corresponds with the amount of protein present in the sample [1–3]. The luciferase assay is highly sensitive, lacks background signal, and is cost-effective [4]. The Aurora kinase A (AURKA) belongs to the Serine/Threonine Aurora kinase family. AURKA plays a key role in mitotic cell division, including chromosomal separation and centrosome maturation. The continuous accumulation, activation, and degradation of AURKA during S-, G2/M-, and at mitotic exit drives the cell cycle division in normal cells [5]. AURKA transcriptionally

Victor M. Bolanos-Garcia (ed.), *Mitotic Exit: Methods and Protocols*, Methods in Molecular Biology, vol. 2874,
https://doi.org/10.1007/978-1-0716-4236-8_12,

downregulates the Tap73 targets p21 (Cipl/WAF1), a Cyclin-dependent kinase (CDKl/CDK2) inhibitor and prevents the G1/S or G2/M transition during DNA damage [6, 7]. There are number of upstream molecules/transcription factors that activate AURKA. For example, NANOG binds to the promoter region of the AURKA gene at multiple sites within 1 kb upstream of the transcription start site (TSS) with maximum enrichment at −757 to −347 bp upstream of TSS [8]. The dual-luciferase reporter assay can be employed to check the activity of the AURKA promoter regulated by NANOG. For this purpose, the AURKA-luc promoter is expressed in cells overexpressing NANOG or suppress the NANOG by RNAi or shRNA.

Principle A dual reporter luciferase assay system involves using two different luciferase enzymes as reporters, typically firefly (*Photinus pyralis*) and Renilla (*Renilla reniformis*) luciferases. The gene of interest is linked to one type of luciferase gene (typically firefly luciferase) in a plasmid and a second plasmid contains a different luciferase gene known as the reporter gene. The plasmids are transfected into cells to express the genes. The luciferase enzyme produced along with the expressed protein can catalyze the reaction to produce light, and the amount of light produced is directly proportional to the amount of gene expressed (Fig. 1).

2 Material

2.1 General Requirements

Plastic ware must be autoclaved. We preferably use RNAse/DNAse-free water. The reagents used should be of molecular biology grade, and attention should be paid to the handling and storing of the reagents. The reagents should be stored according to the manufacturer's instructions. All pipetting setups should be contamination free.

2.2 Cell Culture

1. 0.5, 1.5, and 2.0 mL microcentrifuge tubes and 6, 12, 24, and 96 well plate.
2. Vortex mixer.
3. 15 and 50 mL tubes.
4. Benchtop Centrifuge for tubes.
5. 10 μL, 100 μL, and 1 mL micropipette.
6. Pipette tips.
7. Phosphate buffer saline solution (pH 7.4 10X).
8. Dulbecco's Modified Eagle Medium (DMEM) and Roswell Park Memorial Institute (RPMI)-1640 media.

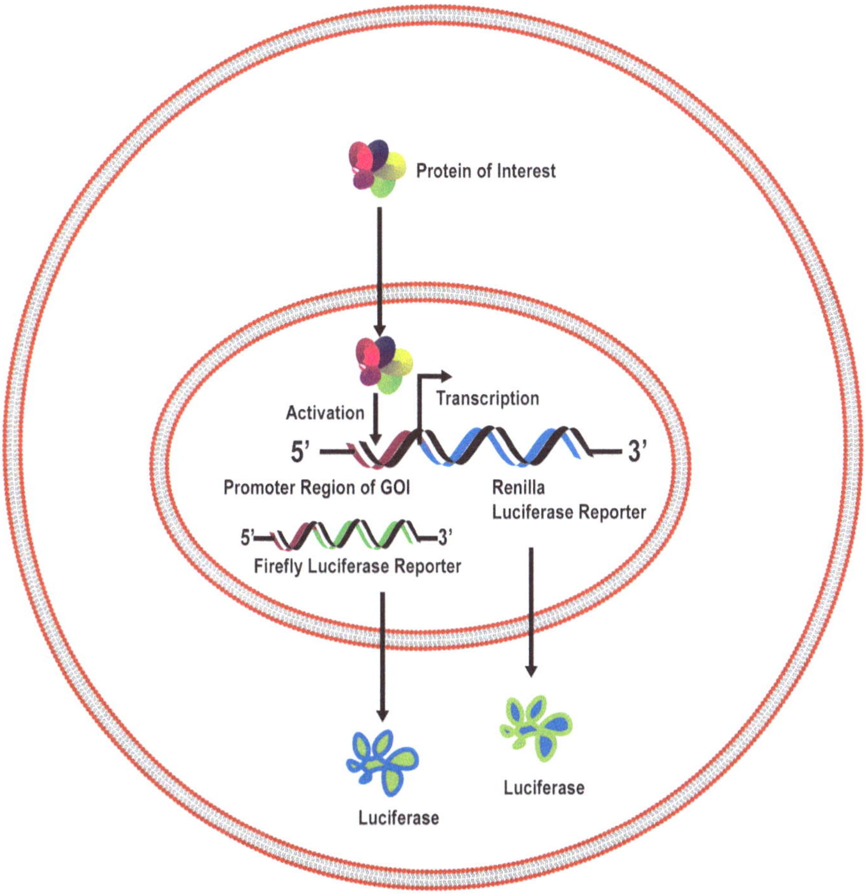

Fig. 1 Principle of Dual-Reporter Luciferase Assay used for promoter activity. The promoter of gene interest is inserted in the plasmid upstream to the luciferase enzyme gene. When the protein interacts with the promoter, it will lead to the expression of the luciferase enzyme, which can be quantified by measuring the intensity of the bioluminescent reaction between enzyme and substrate

9. Trypsin.
10. Fetal Bovine Serum.
11. Antibiotics (Penicillin/Streptomycin).
12. CO_2 Incubator (Thermofisher Scientific, Forma Series II Water Jacket).

2.3 cDNA Construction

1. Primer (AURKA) - 5′-ATGGACCGATCTAAAGAAAA-3′ and 5′ GCCAGTTCCTCCTC AGGATT-3′.
2. Primer (5′ flanking region of AURKA)- 5′- -GGGGGATCCTCACATGAGAGATTAGAGG C-3′ and 5′- -GGGGGATCCCTCTAGCTGTAATAAGTAAC-3′.
3. PCR/Thermocycler (Bio-Rad).

2.4 Plasmid Construction

1. pCDNA3.1-NANOG and pBluescript ii SK (−).
2. Restriction endonucleases *Bam*HI and *Bal*II.
3. T4DNA ligase.
4. c-DNA-Kit.
5. Reporter luciferase plasmid.
6. Qiaprep Miniprep.

2.5 Transfection

1. Lipofectamine 3000 Reagents (Invitrogen).
2. Plasmid DNA (0.5-5 μg/μL stock).
3. Opti-MEM Reduced-Serum Media.

2.6 Reporter Assay

1. 5X Passive Lysis Buffer (Promega).
2. Luciferase Assay Buffer II (Promega).
3. Lyophilized Luciferase Assay Substrate (Promega).
4. Luminometer.

3 Methods

3.1 Plasmid Construction (Fig. 2)

3.1.1 Selection of Plasmid Vector

1. Start by choosing a plasmid vector that contains a multiple cloning site (MCS) upstream of the luciferase gene. This vector should be compatible with the restriction enzymes and have a selectable marker, such as an antibiotic resistance gene.
2. PGL-AURKA hybrid construct was prepared using pGL3-basic vector and pBlue 1486 plasmid.
3. pBlue1486 plasmid is prepared by amplifying the 5′-flanking sequence of the AURKA gene and subcloning it in pBluescript II SK (−) at the BamHI site.
4. The DNA fragment of AURKA containing the 1486 bp sequence is excised from pBlue at the BamHI site and subcloned at the BalII site of the pGL3-basic vector.

3.1.2 PCR Amplification of the Promoter

1. Using genomic DNA (AURKA) as a template, amplify the promoter region of the gene of interest with PCR. The designed primers should include the sequences of restriction enzyme sites on the 5′ ends to facilitate ligation into the vector. The specific sequence of the gene of interest promoter region can usually be found in databases like NCBI.
2. 1.8 kb of the 5′ flanking region was amplified using primer 5′-GGGGGATCCTCACA TGAGAGATTAGAGGC-3′ (f) and 5′-GGGGGATCCCTCTAGCTGTAATAAGTAAC-3′ (b) with the BamHI linker at the 5′ end.
3. The cDNA is synthesized using a cDNA kit and according to the manufacturer's instructions.

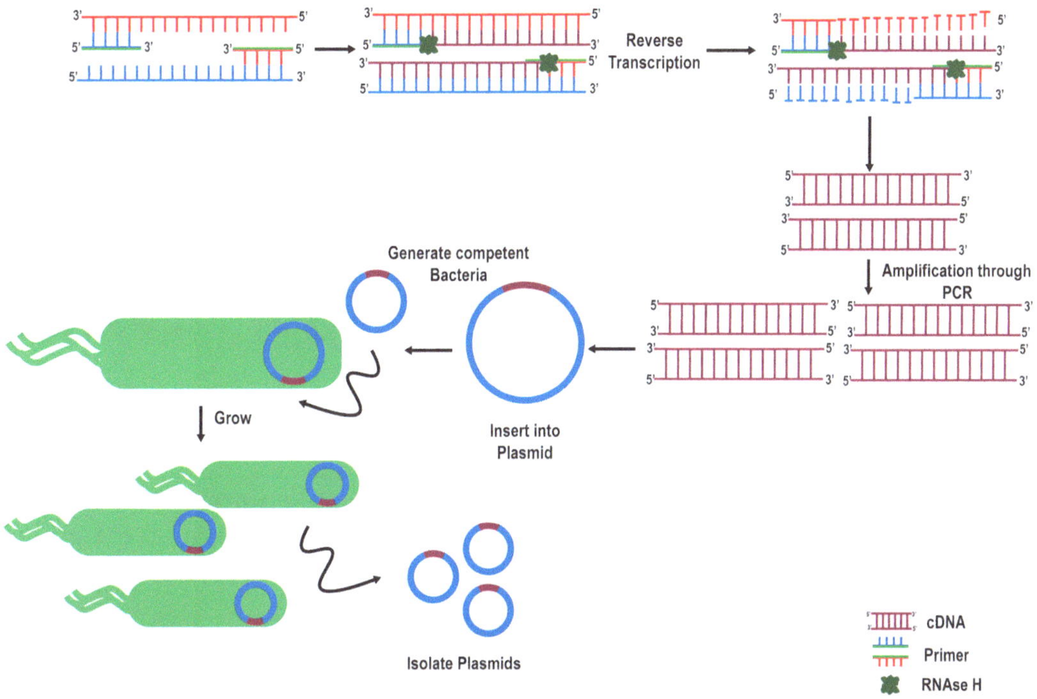

Fig. 2 Steps involved in constructing and isolating plasmid with desired genes. The cDNA of the desired gene is synthesized and inserted inside a vector plasmid. The recombinant plasmid is then transformed inside the bacteria to clone and amplify it. The recombinant plasmids are then harvested from the positive bacterial cells

3.1.3 Digestion

1. Digest the PCR product and the plasmid vector with the appropriate restriction enzymes (BamHI), generating "sticky ends" on both the vector and the insert, allowing them to be joined together.
2. The 5′ flanking sequence of AURKA is subcloned in pBlue plasmid at the BamHI site.
3. To clone the 1486 bp DNA fragment of AURKA in the pGL3 basic, excise the sequence with BamHI enzymes to create a sticky end.
4. The pGL3-basic vector is excised at the BalII site.

3.1.4 Purification

Following digestion, purify the PCR product and the cut vector from the reaction mix using a method such as gel electrophoresis followed by a gel extraction kit.

3.1.5 Ligation

1. Combine the digested vector with the sticky end at the BalII site and DNA with the sticky end at the BamHI site in a ligation reaction, which usually includes a ligase enzyme (T4 DNA ligase) that covalently joins the DNA molecules. This forms a circular piece of DNA that includes the vector containing the luciferase gene and the promoter of the gene of interest.

2. Mix vector and insert in 1:1, 1:3, or 3:1 molar ratio to find the final amount of insert.

$$\frac{\text{ng of vector} \times \text{kb size of insert}}{\text{kb size of vector}} \times \frac{3 \text{ insert}}{1 \text{ vector}} = \text{ng of insert}$$

3. Mix 100 ng of 4.8 pGL3-basic vector DNA with 87.5 ng of 1.4 kb, insert in the solution containing 1 μL of 10X ligase buffer and 0.5u of T4 DNA ligase. Add nuclease-free water to maintain the volume at 10 μL.
4. The reaction mixture should be incubated at room temperature for 3 h.

3.1.6 Transformation

1. Introduce the ligated plasmid into competent bacteria (usually *E. coli*), often via heat shock.
2. Then spread them onto an agar plate containing the appropriate antibiotic. Only bacteria that have taken up the plasmid (and thus the antibiotic resistance gene) will grow.

3.1.7 Verification

1. Pick individual colonies and grow them in liquid media with antibiotics.
2. Extract plasmid DNA from these cultures and verify the correct insertion of the gene of interest promoter through sequencing or restriction digest.

3.1.8 Amplification

Following confirmation of the right construct, a large culture of the bacteria can be grown and plasmid DNA can be extracted using a plasmid prep kit (Qiagen, MD) which involves 3 steps.

Preparation of a Bacterial Lysate

1. Freshly prepared bacterial colonies were picked up and inoculated in 1–5 mL of Luria-Broth medium for 12–16 h at 37 ° C, and the incubated bacterial colony should be subjected to vigorous shaking.
2. Harvest the bacterial culture by centrifuging it at 8500 rpm for 3–5 min at room temperature.
3. Resuspend the bacterial precipitate in 250 μL of Buffer P1 and transfer the suspension into microcentrifuge.
4. By vortexing, ensure the resuspension and complete dissolution of bacterial cells into the buffer.
5. Mix 250 μL of Buffer P2 and thoroughly mix by inverting the tube 4–6 times.
6. Add 350 μL of Buffer N3 to the mixture and mix it by inverting the tube 4–6 times or till the solution becomes cloudy.
7. Centrifuge the mixture for 10 min at 13,000 rpm and discard the pellet.

Binding of DNA onto the Spin Column

1. Add 800 μL of the supernatant to the prep spin column for the selective absorption of DNA in the high salt buffer.
2. Centrifuge for 30–60 s and then discard the flow.
3. Wash the spin column with buffer PE (0.75 mL) and centrifuge again for 30–60 s.
4. Discard the flow and recentrifuged at full speed for 1 min to remove the wash buffer.

Elution of Plasmid DNA

1. For elution of the DNA, put the column in a clean microcentrifuge.
2. Add 50 μL of Buffer EB or water to the center of the spin column, then let it stand for 1 min and centrifuge again for 1 min.

3.1.9 Choosing Plasmid

We prepared the AURKA-Luc reporter plasmid according to the construct developed by Dr. Yoshiaki's lab [9].

1. The plasmid contains a vector backbone of pGL3 basic (Promega) and AURKA promoter.
2. The plasmid is amplified in the DH5α strain of *E. coli* at 37 °C and shows resistance to Ampicillin.

3.2 Preparation of Expression Vector

3.2.1 Selection of Plasmid

Start by choosing the appropriate expression vector. This vector should be compatible with the restriction enzymes and have a selectable marker, such as an antibiotic resistance gene.

1. Select an expression vector to replicate, amplify, and express the NANOG gene in the host cell. pCDNA3.1-NANOG expression vector with vector backbone pcDNA3.1 and hNANOG of 918 insert size were chosen.
2. Centrifuge the vial at 6000 g for 1 min in a refrigerated centrifuge at 4 ° C.
3. After centrifugation, add 100 μL (or desired amount according to the final concentration) of TE buffer to dissolve the DNA. We can also use distilled water instead of TE buffer (**Note 1**).
4. Close the vial and vortex it for 1 min.
5. The prepared DNA is then aliquoted and can be stored or transfected depending on the need.

3.3 Cell Culture

An appropriate cell line (e.g., Huh7 and PC3) overexpressing gene (NANOG) that can activate/suppress promoter of AURKA is selected to check AURKA activity.

1. Maintain the selected cell line according to standard cell culture techniques.
2. The whole process should be carried out in a contamination-free aseptic environment.

3. All equipment used in cell culture should be cleaned with 70% ethanol, and plastic wares must be autoclaved before use.
4. The hood sash should be closed for proper laminar airflow, and cluttering must be avoided inside the hood.
5. Media, supplements, and reagents should be sterilized to prevent microbial contamination.
6. Some reagents will be required to be filter sterilized. Cells should be grown in T-25 to T-75 flask (**Note 2**) in appropriate media (**Note 3**).

3.3.1 Checking Cells

1. Cells should adhere to the bottom of the flask and appear round and plump or elongated in shape and refract light around their membrane.
2. Suspension cells should look round and plump and refract light around their membrane. Some suspension cells may clump.
3. Media should be pinky or orange in color.
4. Cells should be checked daily under a microscope to determine their health and possible contamination.

3.3.2 Dead Cells

1. Cells should be discarded if they detach in large numbers and look shriveled and grainy/dark in color.
2. Cells should be checked for contamination.
3. Cells should be discarded if the media looks cloudy or has turned yellow.
4. Seeding a new batch of cells is better than continuing growing damaged seeded cells.

3.3.3 Subculturing

1. Aseptic technique and sterile conditions are important for subculturing.
2. Refer to the literature on cell lines for an appropriate split ratio. There is frequently a suggested subculture seeding density for suspension cell lines. If a high split ratio is applied, some cells with a moderate growth rate might not grow.
3. To prevent overgrowth, some fast-growing cells could need a high split ratio.
4. Because the seeding density will be too low for the cells to survive, most cells should not be split more than 1:10 in the ratio (**Note 4**).
5. Further, passage numbers should also be recorded and not too high to prevent genetic variations in the cell line.
6. Transfection of plasmid DNA into mammalian cells using Lipofectamine is a common procedure in molecular biology (**Note 5**).

3.4 Cell Seeding

1. Start with a culture of mammalian cells at ~70–90% confluency.
2. Seed the cells into a 24-well plate (or 6-well or 96-well) depending on the experimental setup at a density such that they will be 70–80% confluent at the time of transfection (**Note 6**).

3.5 Transfection

Day 0:

1. Take 1.5 μL Lipofectamine 3000 reagent and 1.5 μg of DNA (reporter and Renilla luciferase plasmid) in two separate tubes in combination with pGL3AURKA, pGL3 empty vectors (control), and sh-NANOG (for inhibiting endogenous NANOG expression.
2. Then add 50 μL of Gibco Opti-MEM Reduced-Serum medium. Vortex the mixtures for 2–3 sec to ensure even mixing.

Day 1:

3. Once the Opti-MEM serum dilutes the reagent and DNA, prepare the master mixture by mixing diluted DNA with diluted Lipofectamine 3000 reagent in 1: 1 ratio.
4. The mixture should be incubated at room temperature for 5 min.
5. Add 100 μL of the mixture to the cells in a complete medium for transfection.
6. Do not add lipid/DNA complex in a few wells to be a negative control.

Day 2:

7. Visualize or analyze the cells in the microscope to check transfection.

3.6 Prepare Cell Lysate

1. The passive lysis buffer provided with the luciferase assay kit is a 5X concentrate. Prepare 1X working concentration of passive lysis buffer by mixing 1-part 5x passive lysis buffer and 4-part distilled water (**Note** 7).
2. Forty-eight hours after transfection, lyse the transfected cells in 1X passive lysis buffer. Before lysing, determine that all the parameters, such as cell confluency (~90% and incubation time), are at the desired state.
3. Remove the media from the plate and gently wash the cells with 1X PBS to remove dead cells and residual media.
4. Remove the PBS from the culture vessels and introduce 1X passive lysis buffer in each well. Ensure that the buffer covers the monolayer of the culture vessel (**Note 8**).

5. To ensure complete coverage of the monolayer of the cells, place the cell culture plate on the rocking platform or shaker. Fix the speed of the shaker at low for gentle shaking (**Note 9**).
6. Rock the plate for 15 min at room temperature.

OR.

7. Homogenous lysates can also be prepared by scraping the cells from the culture vessels **(For Active Lysis).**
8. For this purpose, the media is removed from the cell culture plate, and the cells are washed with PBS.
9. After removing PBS, 1X passive lysis buffer is applied to the culture vessels and cells are harvested immediately after applying passive lysis buffer.
10. In the presence of a passive lysis buffer, the cell culture plate is scraped vigorously by disposable plastic cell lifters.
11. Tilt the plate and scrape the cell in your direction, then transfer the accumulated lysate to a separate tube or vial using the pipette (**Note 10**).

3.7 Prepare the Reporter Assay Substrate

1. Resuspend the lyophilized reporter assay substrate in the reporter assay buffer II to prepare luciferase assay reagent II (LARII) (**Note 11**). The reagent can be stored at −20 °C or −80 °C for months (**Note 12**).
2. Stop & Glo substrates are provided at a 50X concentration. To prepare 1X Stop & Glo substrate, add 1 volume of the substrate provided by the manufacturer to the 50X volume of Stop & Glo buffer. The reagent should be prepared in a glass or siliconized polypropylene tube (**Note 13**).

Standard Protocol

1. *First, add 100 μL of LAR II reagent into the appropriate number of wells or luminometer tubes (NANOG-cDNA/ AURKA-luc co-transfected), (empty vector/ pGL3-AURKA, NANOG-shRNA/AURKA-luc co-transfected) to complete the desired number of reporter assay.*
2. *To remove the background signal generated by the instrument and tube/wells, measure the apparent luminescence activity of non-transfected control cells.*
3. *Prepare 20 μL lysate of non-transfected control cells and add 100 μL of LAR II, then take the reading (**Note 14**).*
4. *Set the program of the luminometer in a 2-second premeasurement delay, followed by a 10-second measurement period for reporter assay.*

5. *Transfer the cell lysates (20 μL) into the wells containing LAR II reagents.*
6. *Mix by pipetting 2 or 3 times. Do not vortex.*
7. *Place the plate in the luminometer and initiate the program for reading (**Note 15**).*
8. *Record the activity measurement of firefly luciferase.*
9. *Add 100 μL of Stop & Glo reagent to each sample well or tube, shake the plate, or vortex the tube to mix the solution (Fig. 3).*
10. *Record the reading of Renilla luciferase activity measurement.*

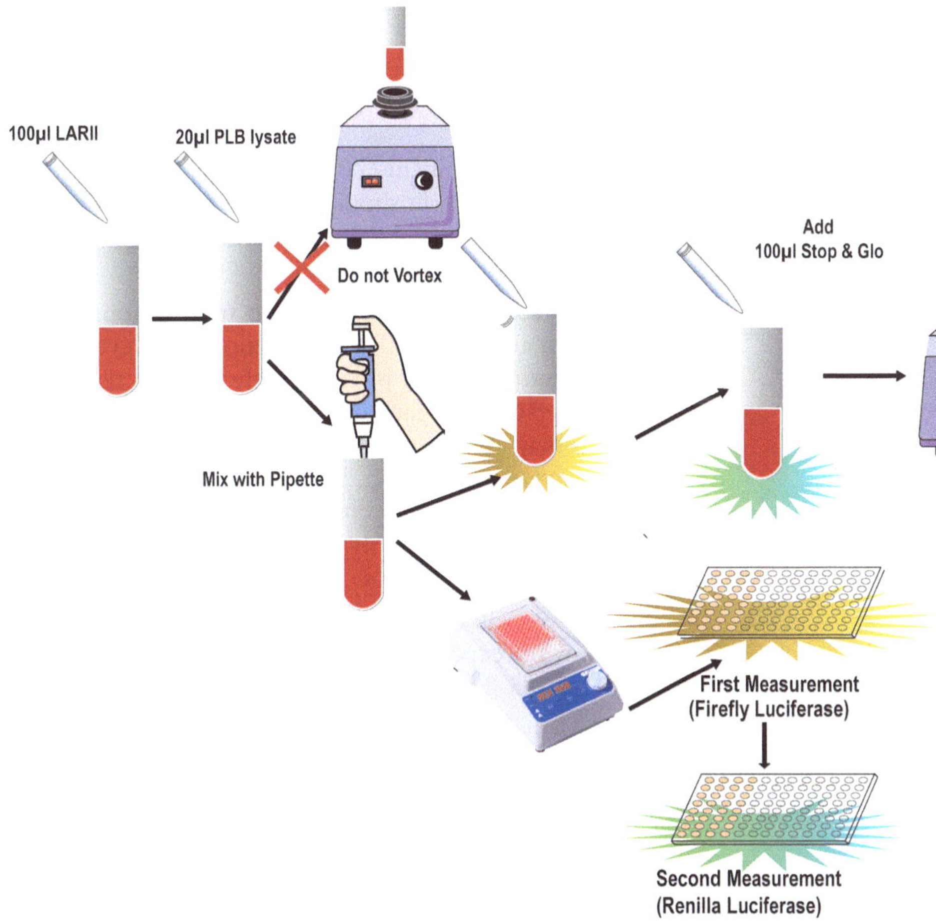

Fig. 3 Steps involved in Dual-Luciferase Reporter Assay. The transfected cells are lysed using passive lysis buffer and cell lysates are transferred into a luminometer tube or well plates. The luciferase substrates are added and their bioluminescence activity is recorded by using a luminometer. PLB—passive lysis buffer; LAR II—Luciferase Assay Reagent

4 Notes

1. DNA can be dissolved in distilled water and TE buffer, and the solvent volume can vary depending upon the desired final concentration. Moreover, if the DNA does not dissolve directly, the solution can be heated at 50 °C for 15 min.
2. Most cell lines do not need special matrixes to grow; however, some cells, particularly primary cells, require special matrixes such as collagen to promote growth, cell attachment, and differentiation. One should carefully read the pertinent literature for more details on the cells to be cultivated.
3. Check the literature provided with the cell line before culturing to determine the appropriate media type, additives, and recommendations. Most cell lines can be cultured in DMEM or RPMI culture media with 10% FBS, 2 mM glutamine, and antibiotics if necessary.
4. 1:2 split should be confluent (70–80%) and ready for the next step in 1 to 2 days. 1:5 split should be ready for a subculture or experiment in 2 to 4 days. 1:10 split should be confluent and ready for experiment in 4 to 6 days.
5. Various other techniques are involved in transfection, such as electroporation and radioactive transfection. However, transfection by electroporation is very costly and needs a well-experienced person. The lack of expertise might result in cell death, which may vary from person to person. Similarly, in the case of radioactive transfection, handling and waste disposal is the major issue. Radioactive emissions might lead to health hazards if not handled properly, so it is better to avoid these techniques as much as possible.
6. The number of cells needed is optimized according to the plates used for the experiment. For 6-, 24-, and 96-well plates, 0.25–1 × 106, 0.5-2 × 105, and 1-4 × 104 cells are required.
7. Diluted passive lysis buffer (1X) can be stored at 4 °C for over a month, but it is advised to calculate the working volume, and according to that, the working concentration of passive lysis buffer should be prepared on the day of the experiment.
8. The recommended volumes of passive lysis buffer for 6-, 12-, 24–48-, and 96-well culture plates are 500 μL, 250 μL, 100 μL, 65 μL, and 20 μL, respectively. Also, overgrown cells should be incubated in the 1X passive lysis buffer for longer as they are more resistant to complete lysis. This will not affect the reporters' activities as the luciferases are stable in the cell lysates prepared in the passive lysis buffer.

9. The cell culture plates should be rocked gently as the high speed might lead to spilling the content out of the cell culture plate.
10. The cells can be frozen and thawed for 1–2 cycles to ensure complete lysis. Also, it is unnecessary to remove the cell debris generated by lysate preparation, but if it is necessary to clear the lysate, then it can be done by centrifuging the lysates for 30 seconds in a refrigerated microcentrifuge.
11. The luciferase assay reagent II should be prepared with the substrate and buffer provided in the kit. The luciferase assay reagent should not be substituted with reagents provided in different luciferase assay systems. Other luciferase assay reagents are not designed for the purpose of dual-luciferase assay.
12. Prepare aliquots of LARII reagent for storage purposes, as repeated freezing-thawing might affect the performance of the reagent.
13. Stop & Glo substrate dilution with the buffer should be done just before the use. Storage of the diluted substrate at room temperature decreases its activity by 8% in 48 h. however, if stored at −20 °C, the substrate's activity is maintained for 15 days. The thawing of the mixture should be at room temperature in a water bath, followed by proper mixing, as thawing generates density and composition gradient.
14. The mammalian cells lack any endogenous luminescence, so any luminescence activity recorded during the reaction is due to the luminometer and plates/tubes etc. The mean of 5–10 readings should be taken to obtain a statistically significant value for the background signals.
15. Suppose the luminometer reading is done using luminometer tubes. In that case, it is advised not to vortex the reagent-lysate mixture, as vortexing may lead to coating the sides of the tubes with the luminescence solution.

5 Calculations

1. Subtract the background reading from the values.
2. Calculate the ratio of firefly-luc to Renilla-luc for each sample.
3. Normalize the data with the control by dividing the normalized value by the average of the control.
4. Use statistical software (GraphPad Prism 9.0) to display the results in mean ± SE.

References

1. Sherf B, Navarro S, Hannah R, Wood K (1996) Dual-luciferase TM reporter assay: an advanced co-reporter technology integrating firefly and Renilla luciferase assays. Promega Notes 57
2. Wood KV (1997) Luciferase assay composition. US5650289A
3. Grentzmann G, Gesteland RF, Atkins JF (2000) Dual-luciferase reporter system. US6143502A
4. Alcaraz-Pérez F, Mulero V, Cayuela ML (2008) Application of the dual-luciferase reporter assay to the analysis of promoter activity in zebrafish embryos. BMC Biotechnol 8:81. https://doi.org/10.1186/1472-6750-8-81
5. Fatma H, Siddique H (2023) AURORA KINASE A and related downstream molecules: A potential network for cancer therapy. Adv Protein Chem Struct Biol 134:115–145. https://doi.org/10.1016/bs.apcsb.2022.10.005
6. Dar AA, Belkhiri A, Ecsedy J, Zaika A, El-Rifai W (2008) Aurora kinase a inhibition leads to p73-dependent apoptosis in p53-deficient cancer cells. Cancer Res 68(21):8998–9004. https://doi.org/10.1158/0008-5472.CAN-08-2658
7. Yang Y, Ding L, Zhou Q, Fen L, Cao Y, Sun J, Zhou X, Liu A (2020) Silencing of AURKA augments the antitumor efficacy of the AURKA inhibitor MLN8237 on neuroblastoma cells. Cancer Cell Int 20:9. https://doi.org/10.1186/s12935-019-1072-y
8. Siddique HR, Feldman DE, Chen CL, Punj V, Tokumitsu H, Machida K (2015) NUMB phosphorylation destabilizes p53 and promotes self-renewal of tumor-initiating cells by a NANOG-dependent mechanism in liver cancer. Hepatology (Baltimore, Md.) 62(5):1466–1479. https://doi.org/10.1002/hep.27987
9. Tanaka M, Ueda A, Kanamori H, Ideguchi H, Yang J, Kitajima S, Ishigatsubo Y (2002) Cell-cycle-dependent regulation of human aurora a transcription is mediated by periodic repression of E4TF1. J Biol Chem 277(12):10719–10726. https://doi.org/10.1074/jbc.M108252200

Chapter 13

Methods for Detecting Phosphorylated Proteins and Observing Intracellular Localization During Mitosis

Miharu Maeda and Kota Saito

Abstract

This protocol describes the detection of phosphorylated proteins within cells and the identification of their intracellular localization, with a particular focus on mitotic cells. While the detection of phosphorylated proteins can be achieved using radioactive labeling with ^{32}Pi, this method presents experimental challenges due to the requirement for radioisotopes. Alternatively, detection using phosphorylation-specific antibodies is a potent method; however, it necessitates the identification of phosphorylation sites and further requires the generation of antibodies targeting these sites, making it effective only for thoroughly analyzed phosphorylations. Here, we outline a convenient method for detecting intracellular phosphorylation using the Phos-tag technique. Additionally, we discuss the establishment of inducible stable cell lines using lentivirus, synchronization methods for cells in the mitotic phase using thymidine and nocodazole, as well as some tips for immunoprecipitation, Western blotting, and immunostaining from mitotic cells.

Key words Mitosis, Phosphorylation, Phos-tag, COPII, Stable cell lines, Immunoprecipitation, Western blotting

1 Introduction

Phosphorylation serves as a covalent post-translational modification that plays a crucial role in regulating the localization, binding specificity and function of the target proteins [1, 2]. Historically, the detection of intracellular phosphorylated proteins has relied on the use of ^{32}Pi labeling of the cells. While this method offers high sensitivity, it requires caution due to the use of isotopes [3]. As a more convenient alternative for broadly measuring the intracellular phosphorylation profile, the Phos-tag technique was developed by Koike's group [4]. Phos-tag is a dinuclear complex that possesses the property of binding to phosphate groups, thus making it usable as a tag for detecting phosphate groups. In order to detect the phosphorylation status of the target protein using Phos-tag, it is essential to purify the target protein through immunoprecipitation, as Phos-tag is a probe for all phosphorylated residues.

Victor M. Bolanos-Garcia (ed.), *Mitotic Exit: Methods and Protocols*, Methods in Molecular Biology, vol. 2874,
https://doi.org/10.1007/978-1-0716-4236-8_13,

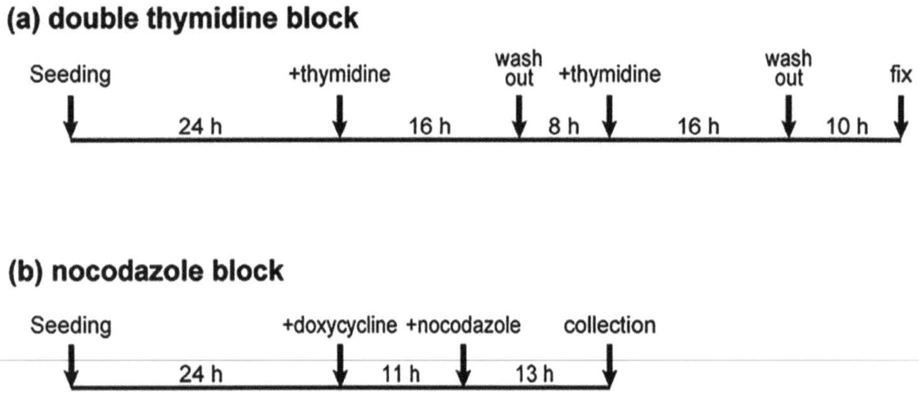

Fig. 1 The time course for synchronizing mitotic phases while simultaneously conducting other manipulations on the cells. (**a**) Double thymidine block with siRNA transfection. (**b**) Nocodazole block with doxycycline induction of inducible stable cell lines

We have previously reported a detailed analysis of the phosphorylation status of TANGO1, a protein localized to the endoplasmic reticulum (ER) exit site, a specialized domain of the ER, during mitotic phase using the Phos-tag biotin method [5]. For this purpose, we made stable inducible cell lines using third generation of lentivirus. In this protocol, we outline the procedure for generating stable cell lines building upon our previous analyses using these cells [5–9].

Furthermore, we introduce the detailed protocol for immunostaining of cells synchronized in the mitotic phase using the double thymidine method (Fig. 1a) to detect ER exit sites, employing cold methanol fixation. We also provide protocols for combining double thymidine synchronization with siRNA-mediated knockdown or nocodazole treatment with doxycycline-induced expression to synchronize cells and induce stable expression strains simultaneously (Fig. 1b). We believe that these protocols will serve as a valuable reference for conducting biochemical and cell biological analyses using cells in the mitotic phase.

2 Materials

2.1 Virus Packaging with 293 T Cells

1. 293 T cells.
2. Polybrene Solution (10 mg/mL).
3. Lipofectamine 2000.
4. Dulbecco's Modified Eagle Medium (DMEM) supplemented with 10%. fetal bovine serum (FBS) (*see* **Note 1**).
5. Opti-MEM.
6. 0.22 μm syringe filter.
7. pSlik-Neo TANGO1S-FLAG and mutants.

8. pRSV-Rev (Addgene).
9. pMDLg/pRRE (Addgene).
10. pMD2.G (Addgene).

2.2 Making Inducible Stable Cell Lines

1. HeLa cells.
2. DMEM+ 10% FBS.
3. G418 solution: G418 sulfate is solved in ddH_2O for 50 mg/ml.
4. 96 well plates.
5. Doxycycline solution: Doxycycline Hydrochloride n-Hydrate is solved in ddH_2O for 1 mg/ml and sterile-filtered.
6. L-alanyl-L-glutamine.

2.3 Double Thymidine Block

1. HeLa cells.
2. 20 μM siRNAs.
3. Lipofectamine RNAi max.
4. Opti-MEM.
5. Thymidine.
6. DMEM+10%FBS.
7. Poly-L-lysine solution: Poly-L-Lysine is solved in ddH_2O for 0.1%(w/v) and sterile-filtered.
8. 12 mm Φ cover slips No.1.
9. 24 well plates.

2.4 Immunofluorescence

1. Cold methanol: Store at −30 °C.
2. Blocking buffer (IF): 5%BSA(w/v)/PBS/0.1%(w/v) Triton X-100/0.1%(w/v) sodium azide filtered with 0.45 μm syringe filter.
3. Primary antibodies diluted in Blocking Buffer.
4. Alexa Fluor conjugated anti-mouse or rabbit or rat IgG.
5. TO-PRO3 or DAPI Solution (1 mg/mL).
6. Aqueous mounting medium.
7. Confocal laser scanning microscopy (Plan Apochromat 63 × / 1.40 NA oil immersion objective lens; LSM 700 or LSM900; Carl Zeiss).

2.5 Nocodazole Block to Collect Cells During Mitosis

1. HeLa cells.
2. Doxycycline solution.
3. Nocodazole solution: Nocodazole is solved in DMSO for 1 mg/mL.
4. DMEM+10% FBS.

2.6 Preparing Cell Lysates

1. Extraction Buffer: 20 mM Tris–HCl (pH 7.4), 100 mM NaCl, 1 mM EDTA, 15 mM NaF, 10 mM $Na_4P_2O_7$, 50 mM Sodium glycerophosphate, 1% (w/v) TritonX-100, Leupeptin, Aprotinin, AEBSF.
2. Pierce Protein Assay Reagent.
3. 4x Laemmli Sample Buffer: 50 mM Tris–HCl (pH 6.8), 10% (w/v) glycerol, 2% (w/v) SDS, 5% (v/v) β-mercaptoethanol, 0.02% (w/v) bromophenol blue.

2.7 Immunoprecipitation

1. Wash Buffer: 20 mM Tris–HCl(pH 7.4), 150 mM NaCl, 0.1% (w/v) TritonX-100.
2. FLAG M2 Agarose Beads.
3. Elution buffer: 200 μg/ml DYKDDDDK peptide in Wash buffer. DYKDDDDK peptide is dissolved in TBS for 5 mg/ml stock solution. Stock solution is then diluted with Wash buffer.
4. 4x Laemmli Sample Buffer: 50 mM Tris–HCl (pH 6.8), 10% (w/v) glycerol, 2% (w/v) SDS, 5% (v/v) β-mercaptoethanol, 0.02% (w/v) bromophenol blue.

2.8 Western Blotting

1. Mini-Protean Tetra Systems.
2. 40%(w/v)Acrylamide/Bis Mixed Solution (37.5:1).
3. 1.5 M Tris–HCl (pH 8.8).
4. 0.5 M Tris–HCl (pH 6.8).
5. 10%(w/v) SDS.
6. 10% (w/v) Ammonium Persulfate (APS).
7. Tetraethylethylenediamine (TEMED).
8. Running Buffer: 25 mM Tris, 192 mM Glycine, 0.1% (w/v) SDS.
9. Molecular weight marker.
10. Blotting Buffer: 15% (v/v) Methanol, 56.2 mM Tris, 38.6 mM Glycine, 0.0375% (w/v) SDS.
11. Methanol.
12. Transblot Turbo Transfer System.
13. Blocking One (Nacalai tesque) or equivalent (5%(w/v)BSA/TBS).
14. Plastic container (6 × 9 cm).
15. Blocking One-P (Nacalai tesque) or equivalent (5%(w/v)BSA/TBS).
16. PVDF membrane.
17. Filter paper.

18. TBS/0.02%(w/v) Tween20.
19. Primary antibody solution: Dilute primary antibody with Blocking buffer. The dilution ratio is adjusted depending on the antibody used.
20. Secondary antibody solution: Dilute secondary antibody (HRP-conjugated IgG) with TBS/0.02%(w/v) Tween20.
21. Plastic container (14 × 10 cm).
22. Chemiluminescent Substrate.
23. CCD imager (LAS4000 mini or ImageQuant 800: cytiva).

2.9 Biotylated Phos-tag Detection

1. Phos-tag biotin (Fuji Film Wako).
2. 10 mM $Zn(NO_3)_2$
3. Streptavidin-conjugated HRP.
4. Amicon Ultra 30 k.
5. TBS/0.02%(w/v) Tween20.
6. Peroxidase Streptavidin.

3 Methods

3.1 Virus Packaging with 293 T Cells

1. Prepare confluent 293 T cell cultures in 10 cm dish. Change into fresh 5 ml of DMEM+10%FBS.
2. Add 3 μg of pSlik-Neo, 3 μg of pRSV-Rev, 3 μg of, 2 μg of pMD2.G vectors into 1.5 mL of Opti-MEM.
3. Add 60 μL of Lipofectamine 2000 to 1.5 mL Opti-MEM. Incubate for 5 min at r.t.
4. Combine 2 and 3 and wait for 15 to 45 min.
5. Add the complexes to 293 T cells in culture dishes.
6. After 4 h incubation, change into fresh medium (DMEM+10% FBS).
7. Culture transfected 293 T cells further for 44 h.
8. Collect culture medium and centrifuge for 1000 rpm at 5 min.
9. Filtrate with 0.22 μm syringe filter.
10. Mix 5 mL of culture medium, 3 mL of fresh DMEM+10%FBS, and 8 μL of Polybrene Solution (10 mg/mL) to make lentivirus medium.

3.2 Making Inducible Stable Cell Lines

1. Before starting, ascertain the ideal concentration of antibiotics for your target cell line (HeLa). To accomplish this, expose the target cells to varying concentrations of G418 and identify the minimal dosage required to eliminate all cells (*see* **Note 2**).
2. Prepare HeLa cells in 20% confluency.

3. Change medium of HeLa cells with lentivirus medium.
4. Culture for 48 h.
5. Change the medium with DMEM+10%FBS containing G418.
6. Change medium with G418 every 2 to 3 days until cells begin to form clumps (*see* **Notes 3** and **4**).
7. Trypsinize the cells and seed them to 96 well plates to achieve 2, 1, 0.5 cells per well.
8. After sufficiently expanding the cells, divide them for checking purposes. Add doxycycline and confirm the expression of the desired construct using appropriate methods such as Western blotting or immunostaining.
9. Titrate the concentration of doxycycline from 1 to 1000 ng/mL for proper induction of desired proteins.

3.3 Double Thymidine Block

1. Making poly-L-lysine coated cover slips (*see* **Note 5**). Dilute Poly-L-lysine solution with ddH_2O to 0.01% (w/v).
2. Place one cover slip into each well of the 24-well plate using a pair of tweezers.
3. Add 200 μL of 0.01% (w/v) poly-L-lysine solution into coverslips.
4. Incubate for 5 min at r.t.
5. Aspirate the solution.
6. Wash with 1 mL of PBS for three times.
7. Let the glass dry.
8. Add 100 μL of Opti-MEM into 24 well plate with poly-L-lysine coated coverslips.
9. Add 0.3 μL of 20 μM siRNA.
10. Add 1 μL of Lipofectamine RNAiMAX.
11. Mix and incubate for more than 5 min.
12. Trypsinize HeLa cells and seed into 24 well plate with 4×10^4 cells.
13. Culture for 24 h (*see* **Note 6**).
14. Change medium to DMEM + 10%FBS with 2.5 mM Thymidine (*see* **Note 7**).
15. Culture for 16 h.
16. Remove medium and wash with DMEM+FBS for twice.
17. Change with fresh DMEM+10%FBS and culture for 8 h.
18. Change medium to DMEM+10%FBS with 2.5 mM Thymidine.
19. Culture for 16 h.
20. Remove medium and wash with DMEM+10%FBS for twice.

21. Change with fresh DMEM+10%FBS and culture for 10 h.
22. Check if about 10% of cells are rounded, indicating the mitotic phase.
23. Process for Immunofluorescence.

3.4 Immunofluorescence

1. Remove medium from 24 well plates with coverslips.
2. Wash with PBS.
3. Add 600 μL of cold methanol.
4. Incubate for 6 min at −30 °C.
5. Wash the coverslips with PBS for twice.
6. Add 250 μL of Blocking buffer.
7. Incubate for 15 min at r.t.
8. Remove Blocking buffer and add 250 μL of primary antibody in Blocking buffer.
9. Incubate for 1 h at r.t.
10. Wash with PBS.
11. Add 250 μL of primary antibody in Blocking buffer.
12. Incubate for 1 h at r.t.
13. Wash with PBS for three times.
14. Add 1000-fold diluted TO-PRO3 in PBS or 10,000-fold diluted DAPI solution in PBS and incubate for 5 min (*see* **Note 8**).
15. Wash with PBS for three times.
16. Mount coverslips onto the slide with mounting medium.
17. Allow it to sit in a dark place for at least 30 minutes before conducting microscopic analysis (*see* **Note 9**).

3.5 Nocodazole Block to Collect Cells During Mitosis

1. Seed 2×10^6 HeLa cells into a 10 cm cell culture dish.
2. Culture for 24 h (*see* **Note 10**).
3. Add doxycycline to reach the appropriate concentration.
4. Culture for 11 h.
5. Change medium with DMEM+FBS containing 100 ng/ml Nocodazole and the appropriate concentration of doxycycline.
6. Culture for 13 h.
7. Check if about 50% of cells are rounded, indicating the mitotic phase.
8. Collect culture medium and centrifuge at 1000 rpm for 5 min at 4 °C. Remove the supernatants.
9. For cell culture dish, add 5 ml of ice-cold PBS and shake-off the mitotic cells by tapping the side of the plate to bench. Collect

the floating cells and transfer them to the centrifuged tube (*see* **Note 11**).

10. Repeat Step 9 for twice (*see* **Note 12**).
11. Centrifuge at 1000 rpm for 5 min at 4 °C.
12. Remove the supernatants (*see* **Note 13**).

3.6 Preparing Cell Lysates

All procedures after extraction should be conducted at 4 °C.

1. Add more than tenfold volumes of Extraction buffer to the collected cells in PBS (*see* **Note 14**).
2. Vortex vigorously.
3. Rotate for 15 min at 4 °C.
4. Centrifuge at 20,000 g for 15 min at 4 °C.
5. Collect the supernatants.
6. (Optional) Protein concentration can be measured and normalized with Protein assay reagents. When measured, add extraction buffer to the concentrated sample and adjust so that the protein concentration becomes constant and proceed to immunoprecipitation.
7. Set aside 30 μL of the supernatant for lysate blotting. Add the same amount of the remaining supernatants to the washed beads. For cell lysate, add 10 μL of 4x Laemmli Sample Buffer and mix. Boil for 5 min.

3.7 Immuno-precipitation

All procedures should be conducted at 4 °C.

1. Prepare beads for immunoprecipitation. Prepare beads of 20 μL bed volume for each sample into one tube (*see* **Note 15**).
2. Add 1 ml of Wash buffer.
3. Vortex.
4. Centrifuge at 6000 rpm for 1 min.
5. Remove the supernatant.
6. Repeat Step 2 to Step 5 for twice more.
7. Add 1 ml of Wash buffer.
8. Vortex.
9. Divide into the number of samples.
10. Centrifuge at 6000 rpm for 1 min.
11. Remove the supernatant.
12. Add lysates into divided beads.
13. Rotate for 2 h at 4 °C.
14. Centrifuge at 6000 rpm for 1 min.
15. Remove the supernatant.

16. Add 1 ml of Wash buffer and vortex.
17. Centrifuge at 6000 rpm for 1 min.
18. Remove the supernatant.
19. Repeat Step 10 to Step 12 for more 4 times.
20. Add 35 μL of Elution buffer.
21. Vortex for 15 min at 4 °C.
22. Centrifuge at 6000 rpm for 1 min.
23. Add 10 μL of 4x Laemmli Sample Buffer into 30 μL of the supernatant and boil for 5 min.

3.8 Western Blotting

1. Prepare the separation gel. Combine the ingredients in the following order.

dH_2O	3.65-x mL
40% (w/v) acrylamide/Bis mixed solution (37.5:1)]	x mL
1.5 M Tris–HCl (pH 8.8)	1.25 mL
10%(w/v) SDS	50 μL
10%(w/v) APS	40 μL
TEMED	10 μL

The amount of Acrylamide/Bis Mixed Solution depends on the final percentage of the gels you would like to make.

2. Pour the gel into the Mini-protean system. Make sure to leave the space for the stacking gel.
3. Layer the top with 1-buthanol.
4. Let it polymerize for 15 min.
5. Remove 1-buthanol and wash with dH_2O.
6. Prepare the stacking gel. Combine the ingredients in the following order.
7. Pour the gel onto the separation gel.
8. Add combs and wait for 15 min to let it polymerize.
9. Clamp gels to the system and fill the tank with running buffer.
10. Apply the samples and molecular weight markers.
11. Begin electrophoresis at 60 V.
12. When the molecular weight marker begins to separate, increase the voltage to 150 V for electrophoresis.
13. Terminate electrophoresis when the dye front has migrated completely.

14. Perform Semi-dry Western blotting with Transblot Turbo Transfer System. Soak filter paper in blotting buffer (*see* **Note 16**).
15. Soak the PVDF membrane in methanol and then place it on top of the filter paper.
16. Place gel on top of membrane.
17. Place filter paper on top of gel.
18. Place items Step 14 through Step 17 on top of the blotter, and use a roller to remove air bubbles.
19. Blot with 12 V for 45 min.
20. Wash the blotted PVDF membrane with TBS.
21. Incubate the blotted membrane for 30 min in Blocking One (*see* **Note 17**) with gentle agitation at r.t.
22. Incubate the blotted membrane for 1 h in primary antibody solution with gentle agitation at r.t. (*see* **Note 18**).
23. Wash the membrane with TBS/0.02%(w/v) Tween20 for 5 min x3 with gentle agitation.
24. Incubate the blotted membrane for 1 h with secondary antibody solution with gentle agitation at r.t.
25. Wash the membrane with TBS/0.02%(w/v) Tween20 for 10 min x3 with gentle agitation.
26. Mix equal volumes of Peroxidase solution and Luminol Reagent from Chemiluminescent HRP substrate into 6x9 container.
27. Soak the membranes into the HRP substrate and place it onto the plastic container.
28. Expose with CCD imagers (*see* **Note 19**).

dH_2O	1.57 mL
40% (w/v) acrylamide/Bis mixed solution (37.5:1)]	0.26 mL
0.5 M Tris–HCl (pH 6.8)	0.625 mL
10%(w/v) SDS	25 μL
10%(w/v) APS	20 μL
TEMED	5 μL

3.9 Biotylated Phos-tag Detection

1. Preparation of Phos-tag solution (see Note 20). Mix 1 μL of 1 mM Phos-tag biotin, 5 μL of 10 mM $Zn(NO_3)_2$, and 1 μL of Streptavidin-conjugated HRP to 479 μL of TBS/0.02%(w/v) Tween20.
2. Incubate for 30 min at r.t.

3. Add mixture into Amicon Ultra 30 k and centrifuge at 10,000 g for 10 min.
4. Collect concentrated phos-tag solution from Amicon Ultra by spin down.
5. Add phos-tag solution into 30 mL of TBS/0.02%(w/v) Tween20.
6. Please refer to **steps 1** to **19** of the previous Subheading 3.8 for creating SDS-PAGE gels, conducting electrophoresis, and performing Western blotting (*see* **Notes 21** and **22**).
7. Wash the blotted PVDF membrane with TBS/0.02%(w/v) Tween20 for 10 min x3.
8. Incubate with Phos-tag solution for 1 h with gentle agitation at r.t.
9. Wash the membrane with TBS/0.02%(w/v) Tween20 for 10 min x3 with gentle agitation.
10. Mix equal volumes of Peroxidase solution and Luminol Reagent from Chemiluminescent HRP substrate into 6x9 cm container.
11. Soak the membranes into the HRP substrate and place it onto the plastic container.
12. Expose with CCD imagers (*see* **Note 23**).

4 Notes

1. It is important to note that antibiotics present in FBS may induce protein expression. In such cases, it is advisable to use tet-system approved FBS.
2. In our condition, G418 selection concentration for HeLa cells is 500 μg/ml.
3. By substituting L-alanyl-L-glutamine for glutamine, you can extend the interval for medium exchange from 2 to 3 days to approximately 1 week.
4. Typically, selection process requires about 10 to 15 days dependent on the infection conditions. Be aware that culture media, culture dishes, pipettes, and other materials may carry the risk of viral infection. Additionally, it is crucial to adhere to regulations regarding the use of third-generation lentivirus.
5. Poly-L-lysine coating of coverslips are crucial, since the cells in the mitotic phase are more prone to detachment compared to cells in interphase. Additionally, it is important to handle the cells gently during the fixation and staining processes.
6. The indicated time frame is suitable for HeLa cells with a doubling time of approximately 24 h. Depending on the

doubling time of the cells used, it is necessary to adjust the duration of time after adding thymidine and the duration of time after removing thymidine.

7. We dissolve thymidine powder directly into DMEM to adjust the final concentration to 2.5 mM.
8. To select and observe cells in the cell division phase based on chromosome morphology, nuclear staining is essential. In this case, DAPI and TO-PRO3 are used for nuclear staining. DAPI is used for visual observation, while TO-PRO3 is used for imaging. If filters suitable for DAPI are available, TO-PRO3 may not be necessary. Since TO-PRO3 has maximal excitation at 671 nm and emission at 661 nm, avoid using dyes with similar wavelengths such as Alexa Fluor 647 when using TO-PRO3 for staining.
9. Cells in the mitotic phase may have a greater thickness compared to cells in interphase, leading to differences in the optimal focal plane for microscopic observation.
10. The duration of nocodazole treatment should be adjusted based on the doubling time of the cells used, with the indicated time frame suitable for HeLa cells with a doubling time of approximately 24 h.
11. Rotate the plate between each tap to complete collection of mitotic cells.
12. Be sure to check the plate under the microscope. If there are few rounded cells present, it indicates successful collection of cells in the mitotic phase.
13. The collected cells can be frozen in liquid nitrogen and stored at −80 °C.
14. In the extraction buffer, in addition to protease inhibitors, phosphatase inhibitors are also added as needed to suppress the dephosphorylation reaction after solubilization.
15. When dispensing beads, cut off the end of yellow chips with scissors. Vortex occasionally to prevent the beads from sinking to the bottom.
16. It is advisable to prepare a plastic case that accommodates the filter paper and PVDF membrane perfectly (6 × 9 cm). Prepare bent clips as shown in the diagram (Fig. 2) in advance and set them in the plastic case. Then, add blotting buffer to the case and place filter paper, methanol-soaked PVDF, gel, and another piece of filter paper into the buffer. Submerge everything in the blotting buffer, then lift the clips to smoothly place the entire assembly onto the blotter. It is better to soak the gel in the blotting buffer for about 10 min before blotting.

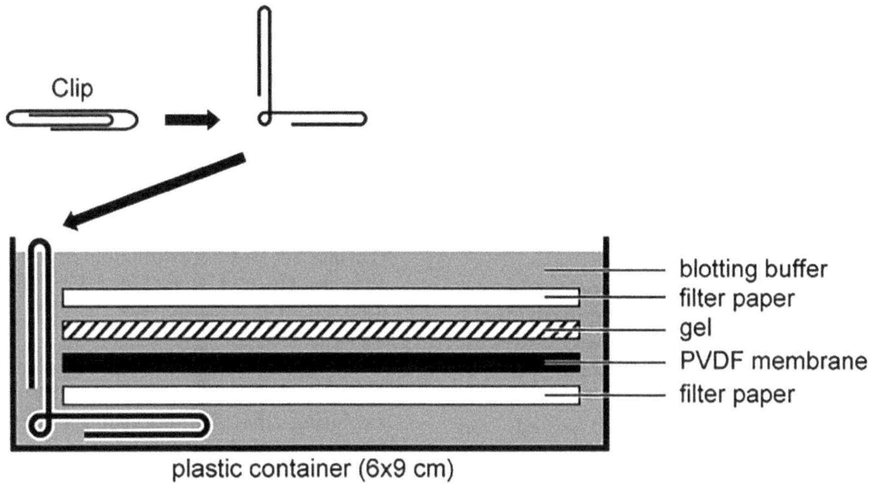

Fig. 2 An overview of Western blotting assembly

17. When performing Western blotting with phosphorylation antibodies, it is crucial that the blocking buffer does not contain phosphorylated proteins. Therefore, instead of using skim milk or regular blocking buffers, it is advisable to use a suitable blocking agent such as Blocking One-P.
18. To confirm whether cells in the mitotic phase have been collected, check the expression of mitotic markers such as cyclin B1.
19. When detecting chemiluminescence, place it on a larger plastic case, such as a 14 × 10 cm case, and observe it using a CCD imager.
20. Since Phos-tag is a probe for all phosphorylated residues, it is necessary to purify the target protein by immunoprecipitation before detecting the phosphorylation status of the target protein using Phos-tag.
21. If the Molecular Weight Marker is detected as a strong signal with Phos-tag and might interfere with signal detection, adjust the amount applied by reducing it to mitigate any potential interference.
22. PVDF membrane is highly recommended for Phos-tag.
23. Phosphorylated proteins often exhibit a band shift toward higher molecular weights.

Acknowledgments

This work was supported by the Japan Society for the Promotion of Science Grants-in-Aid for Scientific Research (23H02430 to K.S. and 22H02760 to M.M.).

References

1. Hunter T (2012) Why nature chose phosphate to modify proteins. Philos Trans R Soc Lond Ser B Biol Sci 367(1602):2513–2516. https://doi.org/10.1098/rstb.2012.0013
2. Delom F, Chevet E (2006) Phosphoprotein analysis: from proteins to proteomes. Proteome Sci 4:15. https://doi.org/10.1186/1477-5956-4-15
3. Sefton BM (1991) Measurement of stoichiometry of protein phosphorylation by biosynthetic labeling. Methods Enzymol 201:245–251. https://doi.org/10.1016/0076-6879(91)01022-t
4. Takeda H et al (2003) Matrix-assisted laser desorption/ionization time-of-flight mass spectrometry of phosphorylated compounds using a novel phosphate capture molecule. Rapid Commun Mass Spectrom 17(18):2075–2081. https://doi.org/10.1002/rcm.1154
5. Maeda M, Komatsu Y, Saito K (2020) Mitotic ER exit site disassembly and reassembly are regulated by the phosphorylation status of TANGO1. Dev Cell 55:237. https://doi.org/10.1016/j.devcel.2020.07.017
6. Maeda M, Saito K, Katada T (2016) Distinct isoform-specific complexes of TANGO1 cooperatively facilitate collagen secretion from the endoplasmic reticulum. Mol Biol Cell 27:2688–2696. https://doi.org/10.1091/mbc.E16-03-0196. Epub 2016 Jul 13
7. Maeda M, Katada T, Saito K (2017) TANGO1 recruits Sec16 to coordinately organize ER exit sites for efficient secretion. J Cell Biol 216(6):1731–1743. https://doi.org/10.1083/jcb.201703084
8. Maeda M et al (2019) COPII proteins exhibit distinct subdomains within each ER exit site for executing their functions. Sci Rep 9:7346. https://doi.org/10.1038/s41598-019-43813-3
9. Shin KJ et al (2006) A single lentiviral vector platform for microRNA-based conditional RNA interference and coordinated transgene expression. Proc Natl Acad Sci USA 103(37):13759–13764. https://doi.org/10.1073/pnas.0606179103

Chapter 14

Studying Mitotic Phosphorylation in *Drosophila*

Éric Bonneil, Myreille Larouche, Virginie Emond-Fraser, Peter Kubiniok, Cristina Mirela Pascariu, Pierre Thibault, and Vincent Archambault

Abstract

Mitosis is largely controlled by the reversible phosphorylation of effector proteins. The addition or removal of phosphate groups alters the activities of these proteins, resulting in changes in chromosome structure, cytoskeletal dynamics, nuclear envelope integrity, and other transformations that must occur as a cell progresses through mitosis. *Drosophila* has been instrumental in the elucidation of the molecular mechanisms of mitosis, which are mostly conserved among animals. In this model system, sophisticated genetic tools can be used to study mitosis in different tissues during development in vivo. *Drosophila* cell culture affords complementary possibilities. In this chapter, we present a phosphoproteomic protocol using *Drosophila* cell culture to identify phosphorylation sites that depend on mitotic kinases and phosphatases. We also provide protocols to dissect the roles of the identified sites in the regulation of protein interactions and localization during mitosis, using *Drosophila* embryos. We emphasize the advantages of the selected methods compared to possible alternatives in *Drosophila* or in other systems.

Key words Phosphorylation, Mitosis, *Drosophila*, Proteomics, Biochemistry, Cell biology, Microscopy

1 Introduction

The intracellular events that mediate mitosis are largely controlled by a subset of kinases and phosphatases that modify specific serine, threonine (and to a lesser extent tyrosine) residues. The mitotic Cyclin-Dependent Kinases (CDKs), Polo-Like Kinases (PLKs) and Aurora kinases are generally more active early in mitosis to promote processes including chromosome condensation, nuclear envelope breakdown and spindle assembly [1–3]. The mitotic Phosphoprotein Phosphatases PP1, PP2A, along with their numerous regulatory subunits, are generally more active late in mitosis to promote chromosome segregation and decondensation, nuclear envelope reassembly, and cytokinesis [4–7]. Precisely how each kinase and phosphatase holoenzyme fulfills its functions is incompletely understood. In particular, the identity of the effector proteins that are

Victor M. Bolanos-Garcia (ed.), *Mitotic Exit: Methods and Protocols*, Methods in Molecular Biology, vol. 2874, https://doi.org/10.1007/978-1-0716-4236-8_14,

phosphorylated or dephosphorylated by each enzyme, and the effects of these modifications on the mitotic process are unclear.

Here, we describe a strategy to better understand how mitosis is regulated by reversible phosphorylation using the *Drosophila* system. The low genetic redundancy in *Drosophila* relative to humans facilitates the dissection of molecular mechanisms using functional genomics and molecular genetics. The mechanisms uncovered in *Drosophila* are mostly conserved in humans and other eukaryotes [8]. We start with the identification of phosphorylation sites in the proteome that are hypo- or hyperphosphorylated after RNAi depletion of a specific kinase or phosphatase, respectively. We then generate transgenic *Drosophila* lines expressing GFP fusions of the phosphoprotein, either WT of with non-phosphorylatable or phosphomimetic mutations. We then use embryos from these flies to affinity purify complexes containing the GFP fusions and identify phosphorylation site-dependent protein interaction. This information is correlated with assessments of differences in protein localization during mitosis using confocal imaging in live embryos.

Phosphoproteomics in *Drosophila* Cells To identify differentially phosphorylated sites on proteins, Stable Isotope Labeling in Cell culture (SILAC) is an effective and popular method that has been previously used in mammalian cells [9]. It relies on the incorporation into proteins of amino acids containing different stable isotopes. Typically, cells are grown in the presence of lysine and arginine containing either heavy isotopes ($^{13}C_6/^{15}N_{2\ \text{and}\ 4}$) in one condition, or light isotopes ($^{12}C_6/^{14}N_{2\ \text{and}\ 4}$) in the other condition. Cells are then mixed in equal ratios and cell lysates are subjected to tryptic digestion, which cleaves after lysine and arginine residues. Phosphopeptides are purified and the relative abundances of heavy vs. light peptides are compared by mass spectrometry. This approach was previously used by our group in *Drosophila* d.mel-2 cells (Schneider 2 cells selected to be cultured without serum). However, we encountered two problems. First, we found that the level of lysine and arginine substitution (heavy replacing light) was low, requiring 9 doublings to reach 90%. In addition, we found that cells proliferated very slowly in the synthetic medium required to enable the amino acids substitution [10]. This is particularly problematic if the goal is to detect mitotic phosphorylation sites because a slower proliferation results in a lower percentage of cells in mitosis. SILAC was used successfully by other groups to study responses to different perturbations outside of mitosis [11, 12]. Instead of SILAC, we use a label-free quantitative approach [10], which is described in this protocol. The kinase or phosphatase subunit of interest can be readily depleted by adding a dsRNA to cells in culture. The phosphoproteome is then mapped quantitatively relative to samples processed in parallel.

Among the sites identified, some may be selected for further studies. Priority may be given to sites that are most highly hyper- or hypophosphorylated, sites found in substrates of particular interest given their known (or unknown) functions in mitosis, or sites that map in domains or regions of the protein that suggest hypothetical regulatory mechanisms. To examine the role of phosphorylation at a given serine or threonine, we substitute these residues by point mutagenesis. An alanine residue mimics constitutively unphosphorylated serine or threonine residues, while aspartic acid or glutamic acid residues mimic constitutive phosphorylation (due to their negative charges) at these sites. Standard methods are available for this mutagenesis and they are not described here.

To confirm the phosphoproteomic result regarding the selected protein, Western blotting may be used to visualize changes in electrophoretic mobility following the depletion of kinase or phosphatase found to act as a putative regulator. Phosphorylation often changes the apparent molecular weight of a protein in SDS-PAGE. Changes in electrophoretic mobility due to mitotic phosphorylation at one site can be detected under standard conditions for some proteins. In these cases, a simple Western blot for a protein expressed in cycling cells can suffice to reveal an upshift of the WT (the phosphorylated form of the WT protein will frequently appear as a distinct band at a higher MW) that is sensitive to the depletion of the upstream kinase or phosphatase, and that is abolished by the mutation. However, such changes can be undetectable or subtle under normal conditions. In this case, the addition of Phos-Tag (Fujifilm) to the gel can be used to enhance the upshift observed. Phos-Tag contains a molecule that binds to phosphate groups and retards the migration of the protein, even under denaturing conditions. In vitro kinase or phosphatase assays using tagged enzymes affinity-purified from *Drosophila* cells and recombinant substrates or synthetic phosphopeptides can further increase confidence that the sites identified are directly modified by the enzyme.

Determination of Phosphorylation Site Function in Mitosis Phosphorylation often regulates the interactions or localization of a protein. For several reasons, better results are generally obtained using young *Drosophila* embryos compared with cells in culture to assess the effects of the phosphorylation site mutations on the interactions and localization of a protein during mitosis. First, *Drosophila* embryos are enriched with mitotic protein complexes because early embryonic development consists of 13 rapid nuclear divisions (approximately every 10 min) in a syncytium [13]. By contrast, *Drosophila* cells in culture divide approximately once per day, and they are notoriously difficult to synchronize or even arrest in mitosis. Second, the use of site-directed transgenic insertions at pre-determined sites in the genome ensures that all transgenic lines and all embryos within each line express the same level of the protein. By contrast, transposon-based methods rely on random genomic integration of the vector, which introduces differences in expression levels between transgenic lines due to local differences in chromatin landscape at the integration sites. Stable *Drosophila* cell lines expressing tagged proteins (WT or mutants) can also be generated and used to probe differences in interactions and localization [10]. However, stable cell lines typically express more variable levels of the expressed protein compared with transgenic embryos.

To generate transgenic *Drosophila* lines, we use plasmid vectors containing attB sites that integrate at attP sites in previously engineered lines [14]. As a result, all animals express the same level of the transgenic protein. Moreover, the WT and mutant forms of the protein are also expressed at the same level. Any difference in levels observed, if any, would reflect differences in protein stability. Constitutive expression in the embryo can be achieved with vectors containing the Ubiquitin promoter (pUb) while inducible expression can be done with vectors containing the UASp promoter [15]. Expression of Gal4 from a driver active in the embryo turns on the expression of the cDNA under control of the UASp (the standard UAS_T does not allow expression in syncytial embryos) [16, 17]. The latter strategy requires additional crossing but it circumvents potential toxicity that can be due to overexpression or expression of a mutant form of the studied protein. In addition, we found that expression from UASp driven by *Maternal α-Tubulin-Gal4-VP16* (Bloomington *Drosophila* Stock Center, line 7062) is stronger than expression from pUb. A series of vectors for the generation of transgenic lines compatible with expression in syncytial embryos was previously described [15]. Plasmid injection and subsequent isolation of transgenic flies or fly lines can be outsourced to companies.

Phosphorylation at a protein interaction site, by changing its electrostatic and shape properties, can disrupt the interaction [18, 19]. In other cases, phosphorylation can enhance the interaction by creating a binding site for a phospho-binding domain such as a FHA or Polo-Box Domain [20, 21]. To assess the potential effect of protein phosphorylation or dephosphorylation on protein interactions during mitosis, we purify the WT, Ala, and Asp mutant forms of the protein in fusion to GFP from young embryos. We previously published in this series a detailed protocol for affinity purifications of GFP-fused proteins expressed from transgenes in *Drosophila* embryos [15]. The GFP tag is bound with picomolar affinity to a nanobody immobilized on agarose beads (GFP-Agarose, Chromotek), allowing very effective purifications of protein complexes. The strategy used here is the same (and will not be copied here), with the addition that the abundances of the peptides from proteins specifically co-purified with the bait are compared between its WT, Ala, and Asp forms. The equal expression levels between all three forms of the protein in embryos is a major advantage when it comes to comparing the protein interaction profiles quantitatively. Various biochemical assays can be used to follow up proteomic results, to confirm and characterize how phosphorylation affects an interaction, the composition of a whole complex, and the molecular functions of its components.

Phosphorylation of a protein often changes its localization. This change may be due to the disruption or enhancement of an interaction of the protein, or to an alteration of its stability. To assess the potential effect of protein phosphorylation or dephosphorylation on protein localization during mitosis, the same transgenes expressing GFP-fused WT, Ala, and Asp forms of the phosphoprotein studied can be used. Another advantage of the embryo is the highly synchronous and reproducible timing of their mitoses within an embryo and between embryos at a given stage, under given conditions [13].

The outcome of the strategy we propose can suggest a potential molecular mechanism where a given kinase or phosphatase enzyme regulates an effector protein at a specific site, allowing changes in the interaction and/or localization of the protein, which may serve a function in mitosis. Investigators may then formulate informed hypotheses for further mechanistic studies.

2 Materials

2.1 *Drosophila* Cell Culture, RNAi, and Harvesting

1. Insect cell medium Express Five SFM (Gibco) or EX-CELL 420 (Sigma).
2. Penicillin/Streptomycin/L-Glutamine 100X.
3. Cell culture flasks, 50 mL, 25 cm^2 (Greiner).
4. dsRNA with sequence identity to mRNA of target protein of interest (see Methods).
5. Standard Tris-Buffer Saline (TBS) pH 7.4 solution.

2.2 Sample Preparation for Mass Spectrometry

1. Acetonitrile.
2. Urea.
3. Tris(2-carboxyethyl)phosphine (TCEP) solution 0.5 M.
4. 2-Chloroacetamide.
5. Ammonium Bicarbonate.
6. Trifluoroacetic acid (TFA).

7. Lactic acid.
8. Digestion buffer for affinity-purified proteins: 10 mM TCEP, 40 mM 2-Chloroacetamide in 50 mM ammonium bicarbonate. Weigh 37.4 mg of 2-chloroacetamide and transfer to a scintillation vial. Add 200 μL of TCEP to the scintillation vial. Weigh 374 mg of ammonium bicarbonate and transfer to the scintillation vial. Add 9.8 mL of HPLC-grade water with a 1000 μL pipette. Store at 4 °C.
9. Cell lysis buffer: 10 mM TCEP 40 mM 2-Chloroacetamide in 8 M urea. Weigh 319 mg of 2-chloroacetamide and transfer to graduated bottle. Add 2 mL of TCEP. Weigh 48 g of urea and transfer to graduated bottle. Complete to 100 mL with HPLC-grade water. Store at 4 °C.
10. Embryo dilution buffer prior to digestion: Weigh 121.1 mg of Tris–HCl and transfer to a scintillation vial. Add 10 mL of HPLC-grade water.
11. Sequencing grade modified Trypsin. Allow the trypsin reconstitution buffer to thaw and add 200 μL to reach a concentration of 0.1 μg/μL of trypsin. Store at −20 °C.
12. Ultra-microspin column (The Nest Group).
13. HLB columns (Waters).
14. Titansphere Phos-TiO Kit (GL Sciences).
15. Desalting washing solution: TFA 0.1%. Place 10 μL of TFA into a scintillation vial and add 9.9 mL of HPLC grade water with a 1000 μL pipette.
16. Conditioning and elution solution: TFA 0.1%, ACN 80%. Place 10 μL of TFA into a scintillation vial and add 1.9 mL of HPLC grade water and 8 mL of ACN with a 1000 μL pipette.
17. Phosphopeptide enrichment buffers: Buffer A 3%TFA, 70% ACN: Mix 1.35 mL of HPLC grade water with 3.5 mL of ACN and 150 μL TFA in a 10 mL Falcon tube. Vortex for 1 min. Buffer B: 3% TFA 70%, ACN 250 mM lactic acid: Mix 1.2 mL of HPLC grade water with 3.5 mL of ACN, 150 μL of TFA and 110 μL of lactic acid in a 10 mL Falcon tube. Vortex for 1 min. Elution Buffer: 1% Ammonium hydroxide pH 10.5: Mix 4.83 mL HPLC grade water with 167 μL Ammonium hydroxide. Vortex for 1 min.

2.3 *Drosophila* Embryo Collection and Preparation for Imaging

1. Grape juice agar plates: Dissolve 25 g of sucrose and 22.5 g of agar in 750 ml of water by boiling. Add 250 mL of grape juice (*see* **Note 1**). Distribute the mix into 60 mm plastic petri dishes, adding 10 mL per dish. Let cool and store at 4 °C in plastic bags.

2. Yeast paste: Mix dry active yeast with water to obtain a paste of the consistency of peanut butter. Store at 4 °C for up to a week.
3. Embryo collection cages, 60 mM (Genesee Scientific).
4. Small paint brush.
5. Bleach solution.
6. Embryo collection mesh baskets (Genesee Scientific).
7. Glass bottom culture dishes, 35 mM (MatTek).
8. Glue solution: fill a screw-cap bottle with clear adhesive tape and heptane. Mix thoroughly to dissolve the glue. Store at room temperature.
9. Voltalef mineral oil (Arkema).

3 Methods

3.1 Identification of Mitotic Phosphorylation Sites in Drosophila Cells

1. For each experimental condition, work in quadruplicates. For each replicate, seed 2×10^7 D-Mel (d.mel-2) cells into 6 mL of Express Five medium supplemented with glutamine, penicillin, and streptomycin in a 25 cm^2 flask (Fig. 1) (*see* **Note 2**). Culture cells at 25 °C.
2. Wait 1 to 2 hours for the cells to adhere properly to the surface.
3. Prepare the transfection mix: 150 μg of double-stranded RNA against the phosphatase or kinase of interest (or non-target) + cell culture medium up to a volume of 2 mL (*see* **Note 3**).
4. Remove the medium from the flasks, taking care not to detach the cells.

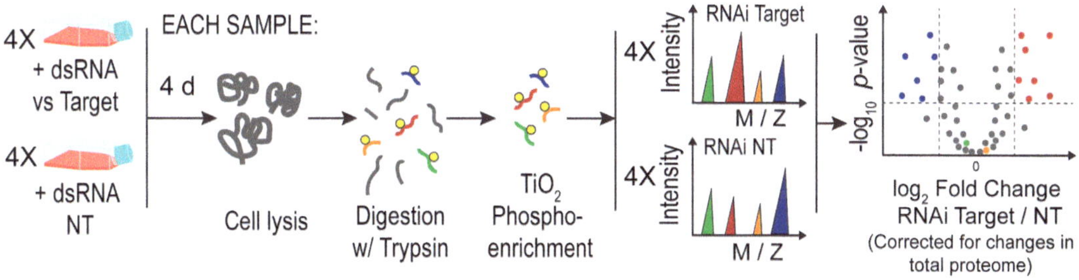

Fig. 1 Experimental scheme for the identification of phosphorylated sites that depend on a mitotic kinase or phosphatase. *Drosophila* cells are transfected with a dsRNA against an enzyme or one of its subunits of interest or with a dsRNA against a non-target (NT) dsRNA in quadruplicates. Four days later, cells are harvested and lysed. Proteins are digested with trypsin and phosphopeptides are enriched using TiO_2 (yellow circles = phosphates). Each sample is analyzed by mass spectrometry. Some peptides are detected as hyperphosphorylated (red) or hypophosphorylated (blue) after depletion of the target (abundances are estimated based on areas under peaks in the spectra). Statistical analysis between quadruplicate samples allows the identification of significant phosphorylation sites

5. Gently add the transfection mixture to the cells immediately.
6. One hour later, add 4 mL of medium into each flask, to a total volume of 6 ml.
7. Four days later, collect the cells by resuspension into the medium using a 1000 μL pipette (*see* **Note 4**). Transfer the cell suspensions into a 15 mL tube.
8. Centrifuge the cells at 150 g for 5 min at 4 °C.
9. Remove the supernatant and resuspend pelleted cells into 1.5 mL of ice-cold TBS buffer pH 7.4. Transfer suspensions to 1.5 mL tubes.
10. Centrifuge the cells at 150 g for 5 min at 4 °C. Remove supernatants (*see* **Note 5**).
11. Add 100 μL of cell lysis buffer and sonicate at room temperature for 40 min. Put the samples on ice and vortex 3 times for 10 s.
12. Centrifuge samples at 1000 g for 10 min.
13. Determine protein concentration using a Bradford assay. Transfer sample volumes corresponding to 2 mg of protein into new 1.5 tubes (*see* **Note 6**).
14. Dilute the samples to 1 M urea with the 100 mM Tris–HCl solution.
15. Vortex samples for 20 min.
16. Add trypsin so that the enzyme/protein mass ratio is 1/100 and incubate at 37 °C on a ThermoMixer overnight.
17. Acidify samples by adding TFA to a final concentration of 1%. Ensure that pH is between 2 and 3 using pH paper. If not add more TFA by 2 μl steps between new measurements.
18. Centrifuge the samples at 13,000 rpm on a tabletop centrifuge (~13,000 g) for 5 min and collect the supernatants in new 1.5 mL tubes.
19. Desalt samples on HLB columns on a vacuum manifold. First, condition each HLB column by adding 1 ml of 0.1% TFA/ACN 50%.
20. Switch on the vacuum pump and open the valve at lowest possible vacuum and gradually increase as needed in order to load the entire volume onto the sorbent bed. Close the valve.
21. Add 1 mL of TFA 0.1% to the HLB columns and repeat **step 20**.
22. Add the acidified samples to the HLB columns and repeat **step 20**.
23. Add 1 mL of TFA 0.1% to the HLB column and repeat **step 20**.
24. Add 1 mL ul of TFA 0.1%, ACN 80% solution and repeat **step 20**.

25. Dry samples in a vacuum concentrator (SpeedVac) without heating.
26. For phosphopeptide enrichment, use the Titansphere Phos-TiO Kit.
27. Reconstitute samples in 200 μL of Buffer B (250 mM lactic acid in 3% TFA and 70% ACN solution), Vortex 10 min and centrifuge at 13,000 rpm for 10 min.
28. Condition TiO_2 microcolumn columns with 100 μl Buffer A: 3% TFA in 70% ACN solution and centrifuge at 3500 rpm for 2 min.
29. Load sample on the microcolumn and centrifuge at 3000 rpm for 2 min.
30. Wash TiO_2 microcolumn column with 100 μL of Buffer B (250 mM lactic acid in 3% TFA in 70% ACN solution) and centrifugate at 3000 rpm for 2 min.
31. Wash TiO_2 microcolumn column with 200 μL of Buffer A (3% TFA in 70% ACN) and centrifuge at 4000 rpm for 2 min.
32. Elute with 200 μl of Elution Buffer (1% Ammonium hydroxide pH 10.5) and centrifuge at 3000 rpm for 2 min.
33. Repeat step 32 twice.
34. Dry samples in a vacuum concentrator (SpeedVac) without heating.
35. Store samples at −30 °C.
36. Prior to liquid chromatography and mass spectrometry (LC-MS) analysis, samples are reconstituted in formic acid 4% and place in a sonication bath for 10 min. Samples are vortexed 15 min in a table vortex and centrifugated at 6000 g for 5 min. They then are placed in the LC autosampler.
37. Samples are then analyzed by LC-MS (*see* **Note** 7).
38. After database search and protein quantitation, phosphopeptide intensities are normalized with the corresponding protein intensities. Phosphopeptides which cannot be matched with a corresponding protein intensity have their intensities normalized in each sample with the median of all protein intensities. Data analysis is then performed with the DEP2 package in R. Phosphopeptides that have not been quantified in all replicates of at least one condition are filtered out. Variance stabilization normalization is performed on all samples. Imputation of missing values is done with Random Forest. Fold Change and p-values are calculated.
39. Generate a figure presenting the results in the form of a volcano plot (Fig. 1).

3.2 Identification of Phosphorylation-Dependent Protein Interactions in Embryos

A statistical analysis is required to conclude with high confidence that the differences in protein interactions are attributable to the phosphorylation site(s) identified. To this end, the following procedure should be performed in triplicate, as we have previously done [10].

1. Proceed to GFP-affinity purifications from embryos expressing GFP fusions of the phosphoprotein of interest in 3 variants: WT, a Ser/Thr → Ala mutant and a Ser/Thr → Asp mutant, and from embryos expressing GFP alone (Fig. 2a). Please refer to the detailed procedure for this purification previously published in this series [15] (*see* **Note 8**). At the end of the procedure, samples will be in the form of dry samples pelleted in 1.5 ml tubes.
2. Freshly prepare the reduction alkylation buffer: 10 mM TCEP, 40 mM 2-Chloroacetamide in 50 mM ammonium

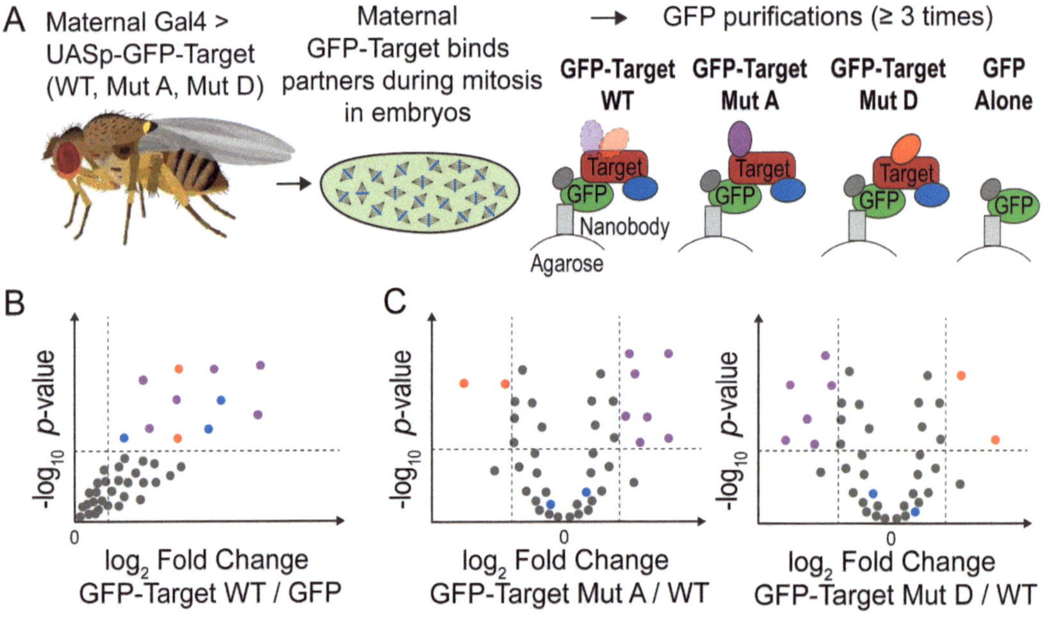

Fig. 2 Experimental scheme for the identification of protein interactors that are sensitive to the mitotic phosphorylation site(s) identified in a target. A. Transgenic *Drosophila* lines are generated for the maternal expression of the GFP-fused phosphoprotein of interest (GFP-Target), either WT, phospho-null (Ser/Thr → Ala; Mut (**a**) or phosphomimetic (Ser/Thr → Asp; Mut D). Early, syncytial embryos are collected and submitted to GFP-affinity purifications using agarose coupled to anti-GFP nanobodies (performed at least three times). Proteins that interact with the GFP-Target variants (or with GFP alone) are co-purified and identified by mass spectrometry. (**b**) Identification of proteins that are specifically and significantly co-purified (colored points) with the GFP-Target compared with GFP alone. (**c**) Identification of proteins that are specifically and significantly enriched or depleted with the GFP-Target Mut A or Mut D compared with GFP-Target WT. Purple and orange points: proteins whose interaction is inhibited or induced by phosphorylation of the target, respectively (see cartoons in Panel B)

bicarbonate. The volume of the prepared reduction/alkylation buffer should be enough to provide 200 µL per sample.

3. Add 200 µL of reduction alkylation buffer to each tube containing the dry protein samples. Use manual pipetting and/or sonication to break up pellets.
4. Vortex 1 h. at maximum speed on a benchtop vortex at room temperature.
5. Agitate 1 h. at 500 rpm at 37 °C on a ThermoMixer.
6. Prepare/thaw trypsin 0.1 µg/µL. Do not vortex.
7. Add 1 µg of trypsin to each tube.
8. Incubate overnight at 500 rpm at 37 °C on a ThermoMixer.
9. Acidify the peptide mixture by adding TFA to a final concentration of 1% (*see* **Note 9**).
10. Vortex 1 min at maximum speed.
11. Centrifuge the samples at 1000 rpm on a tabletop centrifuge (~100 g) for 1 min.
12. Condition each ultra-microspin column by adding 100 µl of TFA 0.1%, ACN 80% and centrifuge at 1000 rpm for 1 min.
13. Add 100 µL of TFA 0.1% to the ultra-microspin column and centrifuge at 1000 rpm for 1 min.
14. Add the acidified sample (~ 200uL) to the ultra-microspin column and centrifuge at 1000 rpm for 1 min.
15. Add 100 µl of TFA 0.1%, ACN 80% and centrifuge at 1000 rpm for 1 min.
16. Repeat **step 15**.
17. Dry the collected 200 µl of eluted peptides in a vacuum concentrator (SpeedVac).
18. Samples are then analyzed by LC-MS (*see* **Note 7**).
19. After database search and protein quantitation, data analysis is then performed with the DEP2 package in R. Proteins that have not been quantified in all replicates of at least one condition are filtered out. Variance stabilization normalization is performed on all samples. Imputation of missing values is done with Random Forest. Fold Change and p-values are calculated.
20. Generate volcano plots presenting proteins interacting specifically with the GFP-fused WT protein compared with GFP alone (Fig. 2b) and with mutant vs. WT versions of the GFP-fused protein (Fig. 2c) in the form of volcano plots.

3.3 Determination of Phosphorylation-Dependent Protein Localization During Mitosis in Embryos

To visualize changes in phosphoprotein localization during the mitotic cell cycle, embryos expressing GFP fusions of the phosphoprotein of interest in three variants: WT, a Ser/Thr → Ala mutant, and a Ser/Thr → Asp mutant are used. Conveniently, embryos expressing the same transgenes can be used for protein purifications (*see* Subheading 3.2) and imaging. However, additional spatial and timing information in mitosis is obtained by generating embryos co-expressing a second fluorescent marker protein in another channel. For example, Histone 2A-RFP can be used to mark chromosomes, and allow visualization of anaphase onset as a reference time for the events of nuclear envelope reformation after mitosis (Fig. 3a).

1. Insert a few hundred female flies of the desired genotypes together with any males in small embryo collection cages fitted onto 60 mM-wide grape juice agar plates with a spot of yeast paste added in the middle. Place the cage standing with the plate down in an incubator at 25 °C with 50% humidity.
2. Embryos can be collected starting on the next day. Replace the old grape juice agar plate with a new one with a spot of yeast paste added in the middle (*see* **Note 10**). Discard the old plate.
3. One to two hours later, replace the plate again and collect the embryos on the exit plate: dip a small paint brush in water and rub it on the embryos to dislodge them from the agar. Place embryos in a small collection mesh basket immersed half-way in water.
4. Remove the basket from water and remove excess water by touching the bottom with a paper towel. Place the basket

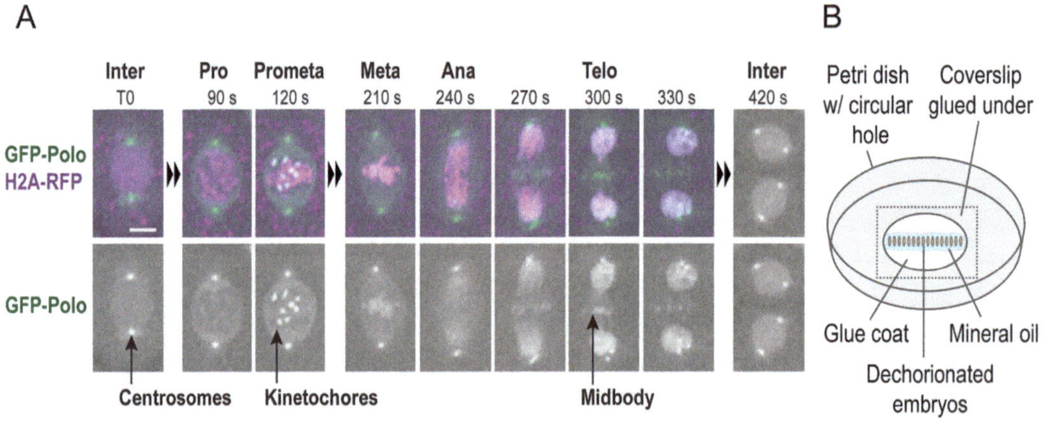

Fig. 3 Live imaging of mitosis in a *Drosophila* embryo. (**a**) Example of imaging in an embryo expressing GFP-Polo (green) and H2A-RFP (magenta). Inter: Interphase; Pro: Prophase; Prometa: Prometaphase; Meta: Metaphase; Ana: Anaphase; Telo: Telophase. Discrete localizations of GFP-Polo at mitotic structures at specific phases and times are indicated. Scale bar: 5 μm. (**b**) Montage of embryos in an imaging dish as explained in the text

immersed half-way in 50% bleach to remove the embryonic chorion. Agitate embryos gently with the paint brush under a binocular stereomicroscope (see **Note 11**).

5. Remove the basket from the bleach and place it in water to rinse embryos. Agitate embryos gently with the paint brush.
6. Remove the basket from water and dry its bottom with a paper towel.
7. Using the brush, place embryos one by one onto the glue-coated coverslip of an imaging dish. For subsequent imaging, it is convenient to place embryos in a line perpendicularly to their anterior-posterior axis. Leave approximately one embryo thickness of spacing between embryos (Fig. 3b).
8. Cover embryos minimally with halocarbon oil using a pipette with a cut tip.
9. Install the montage on a confocal microscope for live imaging (see **Note 12**). Select an oil objective with a magnification of at least 40X. Use the bright field to find the focus on embryos.
10. Position the objective on the first embryo of the line and examine green fluorescence under the 488 nm laser. Look for an embryo at the desired developmental stage where the GFP-fused protein is visible. Mitotic nuclei become visible around cycle 9, as they approach the surface (see **Note 13**).
11. Adjust exposure times for each fluorophore, z-steps, and the interval between time points to acquire optimal data in the mitotic stages of interest (see **Note 14**). If more than one cycle is to be recorded, consider that nuclei continue to get closer to the cortex until cycle 13 when selecting z-steps. A total of 20 z-steps interspaced by 1 μm and time intervals of 10 s may be suitable in several cases.
12. Image several embryos expressing each of the 3 GFP-fused phosphoprotein variants (WT, Ala, Asp) at similar mitotic cycle numbers (see **Note 15**).
13. From the acquired movies, use image analysis software to quantify fluorescence intensity at sites of interest as a function of time relative to a reference time point in mitosis. Obtain measurements from several embryos at the same mitotic cycle number for each genotype. Use statistical analysis to determine if any difference of interest can be attributed to the alteration of the phosphorylation site(s).

4 Notes

1. Artificial grapefruit drinks are suitable.
2. The EX-CELL 420 (Sigma) medium also works well.

3. The sequences of the dsRNAs should be chosen to avoid off-target effects. We use the SnapDragon design tool available online (https://fgr.hms.harvard.edu/snapdragon). dsRNAs can be generated from PCR amplicons using a T7 RiboMAX kit (Promega). We use dsRNA derived from the bacterial kanamycin resistance gene as a non-target control. Other non-target sequences are possible, such as GFP. We previously used transfection reagent to facilitate the entry of the dsRNA into cells, but we found that it is not necessary.
4. The culture time after transfection may be varied to obtain the desired depletion level of the protein and phenotypic penetrance. The complete depletion of proteins essential for mitosis will result in proliferation arrest and/or cell death. In this case, reduced amounts of dsRNA or shorter culture times may be used.
5. At this stage, cell pellets may be flash-frozen by dipping the tubes in liquid nitrogen and stored at −80 °C.
6. For an efficient phosphopeptide enrichment, the starting material should yield at least 1–2 mg of protein. The starting number of cells used has been adjusted accordingly in this protocol, based on our experience. Nevertheless, we recommend that investigators conduct a preliminary test to verify the protein content obtained under their experimental conditions.
7. We currently use a Neo Vanquish nano-liquid chromatography system coupled to an Orbitrap Ascend Tribrid Mass Spectrometer.
8. A starting mass of 300 mg of embryos per genotype may be sufficient, but the amount of starting material may be determined based on expression levels of the GFP fusion proteins. Adjust the volume of lysis buffer used in proportion to the starting mass of embryos.
9. The resulting pH is approximately 3.
10. Pre-warm the plates at 25 °C to avoid stressing the flies and their embryos.
11. A standard fly-pushing stereomicroscope is ideal. When embryo appendages have fallen off (after approximately 2 min), immediately remove the basket and return it to water. Rinse embryos by agitation with the brush.
12. We use a spinning-disc confocal system (Yokogawa CSU-X1 5000) mounted on a fluorescence microscope (Zeiss Axio Observer Z1) using an Axiocam 506 mono camera (Zeiss), 63× oil objective (NA 1.4), and ZEN software.
13. If the embryo is too young or unfertilized, only autofluorescent, spherical yolk particles are visible. Dissipation of these

particles is an indication that development is underway and that mitotic nuclei are approaching the surface.

14. Optimize the settings to obtain the best balance between image quality, spatial resolution, and time resolution. The time point interval should allow the collection of sufficient information to uncover differences during the mitotic stages of interest.
15. The cycle number can be determined from nuclear density or by counting down from cycle 13 (in embryos that were imaged till that stage).

Acknowledgments

This work was funded by a Project Grant from the Canadian Institutes of Health Research (CIHR) to V.A. (grant no. 175132). V.E.F. was a recipient of studentships from the CIHR and from the Fonds de Recherche du Québec–Nature et Technologies. M.L. received a studentship from the Fonds de Recherche du Québec–Santé. P.K. was a recipient of the Vanier scholarship.

References

1. Morgan DO (1997) Cyclin-dependent kinases: engines, clocks, and microprocessors. Annu Rev Cell Dev Biol 13:261–291
2. Archambault V, Glover DM (2009) Polo-like kinases: conservation and divergence in their functions and regulation. Nat Rev Mol Cell Biol 10(4):265–275
3. Carmena M, Ruchaud S, Earnshaw WC (2009) Making the auroras glow: regulation of Aurora a and B kinase function by interacting proteins. Curr Opin Cell Biol 21(6):796–805
4. Chen F, Archambault V, Kar A, Lio P, D'Avino PP, Sinka R et al (2007) Multiple protein phosphatases are required for mitosis in drosophila. Curr Biol 17(4):293–303
5. Moura M, Conde C (2019) Phosphatases in mitosis: roles and regulation. Biomol Ther 9(2)
6. Archambault V, Li J, Emond-Fraser V, Larouche M (2022) Dephosphorylation in nuclear reassembly after mitosis. Front Cell Dev Biol 10:1012768
7. Holder J, Poser E, Barr FA (2019) Getting out of mitosis: spatial and temporal control of mitotic exit and cytokinesis by PP1 and PP2A. FEBS Lett 593(20):2908–2924
8. Rubin GM, Yandell MD, Wortman JR, Gabor Miklos GL, Nelson CR, Hariharan IK et al (2000) Comparative genomics of the eukaryotes. Science 287(5461):2204–2215
9. Hoedt E, Zhang G, Neubert TA (2019) Stable isotope labeling by amino acids in cell culture (SILAC) for quantitative proteomics. Adv Exp Med Biol 1140:531–539
10. Emond-Fraser V, Larouche M, Kubiniok P, Bonneil E, Li J, Bourouh M et al (2023) Identification of PP2A-B55 targets uncovers regulation of emerin during nuclear envelope reassembly in drosophila. Open Biol 13(7): 230104
11. Cuomo A, Bonaldi T (2010) Systems biology "on-the-fly": SILAC-based quantitative proteomics and RNAi approach in Drosophila melanogaster. Methods Mol Biol 662:59–78
12. Sap KA, Bezstarosti K, Dekkers DH, van den Hout M, van Ijcken W, Rijkers E et al (2014) Global quantitative proteomics reveals novel factors in the ecdysone signaling pathway in Drosophila melanogaster. Proteomics 15(4): 725–738
13. Foe VE, Odell GM, Edgar BA (1993) Mitosis and morphogenesis in the drosophila embryo: point and counterpoint. In: Bate M, Martinez Arias A (eds) The development of Drosophila melanogaster. Cold Spring Harbor Laboratory Press, New York, p 746

14. Groth AC, Fish M, Nusse R, Calos MP (2004) Construction of transgenic drosophila by using the site-specific integrase from phage phiC31. Genetics 166(4):1775–1782
15. Lipinszki Z, Wang P, Grant R, Lindon C, Dzhindzhev NS, D'Avino PP et al (2014) Affinity purification of protein complexes from drosophila embryos in cell cycle studies. Methods Mol Biol 1170:571–588
16. Rorth P (1998) Gal4 in the drosophila female germline. Mech Dev 78(1–2):113–118
17. Brand AH, Perrimon N (1993) Targeted gene expression as a means of altering cell fates and generating dominant phenotypes. Development 118(2):401–415
18. Nishi H, Hashimoto K, Panchenko AR (2011) Phosphorylation in protein-protein binding: effect on stability and function. Structure 19(12):1807–1815
19. Nishi H, Shaytan A, Panchenko AR (2014) Physicochemical mechanisms of protein regulation by phosphorylation. Front Genet 5:270
20. Park JE, Soung NK, Johmura Y, Kang YH, Liao C, Lee KH et al (2010) Polo-box domain: a versatile mediator of polo-like kinase function. Cell Mol Life Sci 67(12):1957–1970
21. Almawi AW, Matthews LA, Guarne A (2017) FHA domains: Phosphopeptide binding and beyond. Prog Biophys Mol Biol 127:105–110

Chapter 15

Inducing Exit from Mitosis by Inactivating Cdk1/Cyclin B in Metaphase-Arrested Cells: A Tool to Study Mitotic Exit and Reestablishment of Interphase

James R. Paulson

Abstract

Treatments which inhibit or inactivate Cdk1/cyclin B in metaphase-arrested mammalian cells and budding yeast are described. These treatments induce the cells to exit mitosis and return to interphase, though without chromosome segregation or cytokinesis, and they provide the basis for a method to identify enzymes or other proteins which act "downstream" from Cdk1 inactivation and to elucidate the roles of those proteins in mitotic exit. In this method, inactivation of Cdk1 is combined with inhibition or inactivation of a protein of interest and the effects are observed. This approach should be particularly useful for determining which protein phosphatases are involved in the transition from mitosis to G1-phase and for identifying their substrates.

Key words Cdk1/Cyclin B, Protein kinase inhibitors, Cell cycle, Mitotic exit, HeLa, FT210, *Saccharomyces cerevisiae*, 1NM-PP1

1 Introduction

Mitosis, the process of cell division in eukaryotic cells, has two major stages. The first is the onset of mitosis, in which a cell passes from G2-phase to metaphase. This is triggered by the activation of the protein kinase Cdk1/cyclin B, also known as MPF (M-phase Promoting Factor) [1]. In this transition, many proteins are phosphorylated (by Cdk1 or secondary protein kinases) and the dramatic changes that characterize mitosis take place. The second stage is the completion of mitosis, in which the cell passes from metaphase to G1-phase. This is signaled by inactivation of Cdk1/cyclin B [2], which results from the attachment of all the kinetochores to spindle microtubules and the consequent activation of the Anaphase Promoting Factor (APC). The APC is a ubiquitin ligase which marks cyclin B and other proteins for destruction by the proteosome [3]. Cyclin B destruction inactivates Cdk1 leading to

Victor M. Bolanos-Garcia (ed.), *Mitotic Exit: Methods and Protocols*, Methods in Molecular Biology, vol. 2874,
https://doi.org/10.1007/978-1-0716-4236-8_15,

protein dephosphorylation; reassembly of the nuclear lamina, membrane, and pore complexes; chromosome decondensation; resumption of gene transcription; and in general reestablishment of the interphase state.

This chapter describes methods for artificial inactivation of Cdk1 in metaphase-arrested cells. As reviewed in [2], inactivating Cdk1 in metaphase-arrested cells is sufficient to initiate exit from mitosis and reestablishment of interphase, though without chromosome segregation or cytokinesis. These methods provide tools for studying events "downstream" from Cdk1 inactivation in mitotic exit. To determine whether an enzyme or other protein is involved in exit from mitosis, Cdk1 inactivation is combined with inhibition or inactivation of the protein of interest and the effects are observed. For example, Paulson et al. [4] induced metaphase-arrested HeLa cells to exit mitosis by treating with staurosporine, an inhibitor of Cdk1 and other protein kinases. The effects of various concentrations of the protein phosphatase inhibitors calyculin A and okadaic acid indicated that nuclear reassembly and histone H1 dephosphorylation required Protein Phosphatase 1 (PPase 1). Wu et al. [5] treated interphase Xenopus egg extracts with excess non-degradable cyclin B to arrest them in a meiotic metaphase state. They then inhibited Cdk1 with the kinase inhibitor roscovitine and used phosphatase inhibitors to study the role of PPase 1 during return to an interphase state. Keaton et al. [6] studied budding yeast carrying the *cdc28-as1* allele which makes Cdc28 (Cdk1) sensitive to inhibition by the ATP analog 1NM-PP1 [7]. Treatment of nocodazole-arrested *cdc28-as1* cells with 1NM-PP1 causes them to leave mitosis and enter G1-phase as judged by rebudding, and, when the cells are also exposed to α-factor, "shmooing" (development of mating projections). However, if PPase 1 (Glc7 in *S. cerevisiae*) is inactivated by a cold-sensitive mutation, the cells do not exit mitosis. This shows that PPase 1 is required "downstream" from Cdk1 inactivation during exit from mitosis in budding yeast [6].

The methods described here to artificially inactivate Cdk1 in metaphase-arrested cells could also be useful for identifying the substrates which protein phosphatases act on during mitotic exit. One could inhibit or inactivate the protein phosphatase of interest, then inactivate Cdk1 and analyze which proteins become dephosphorylated and which do not. Mass spectrometry could be used to broadly examine the phosphoproteome, as has been done with yeast [8]. Alternatively, ^{32}P-labeling, western immunoblotting and other techniques could be used to check for dephosphorylation of specific proteins.

The protocols described below are those that we have used most often. However, a variety of other methods and systems have been used to explore Cdk1 inactivation (see [2, 6] for references). The basic approach—inducing metaphase-arrested cells to exit mitosis—is potentially applicable to any dividing cells, provided that one has a way to arrest the cells in metaphase of mitosis and a way to inhibit or inactivate Cdk1 in the cells.

2 Materials

2.1 Chemicals and Solutions

1. 100 mM thymidine in 0.9% NaCl: Filter sterilize, store at 4 °C.
2. Dimethyl sulfoxide (DMSO): Store at room temperature over 4 Å molecular sieve.
3. 5 mg/mL nocodazole in DMSO: Store at −20 °C.
4. 30 mg/mL benomyl in DMSO: Store at −20 °C.
5. 50 mM roscovitine free base (mol. wt. 354.45) in DMSO: Store at −20 °C.
6. 1 mM staurosporine free base (mol. wt. 466.53) in DMSO: Store at −20 °C.
7. 25 mM ATP analog 1NM-PP1 (4-amino-1-*tert*-butyl-3-(1′--naphthylmethyl) pyrazolo[3,4-d] pyrimidine, mol. wt. 331.42) in DMSO: Store at −20 °C.
8. 0.9% NaCl: Sterilize by autoclaving.
9. Phosphate-buffered saline (PBS): Dissolve 0.21 g KH_2PO_4, 0.726 g Na_2HPO_4, and 9 g NaCl per liter of solution. Sterilize by autoclaving.
10. 50 mM sodium orthovanadate ($NaVO_3$, mol. wt. 121.93): Add 61 mg to 10 mL culture medium, stir and warm to dissolve. Use fresh.
11. 50 mM menadione (vitamin K3) (mol. wt. 172.2) in DMSO: Store at −20 °C.
12. 10 mM α-Factor in 0.1 M sodium acetate, pH 5.2 (Zymo Research).
13. 10 mg/mL Hoechst 33342 (bisbenzimide Hoechst 33342 trihydrochloride) stock solution in water. Store dark at room temperature.
14. 20 μg/mL Hoechst 33342: 500-fold dilution in water of the 10 mg/mL stock solution: Store dark at room temperature.
15. Fixative: 3:1 methanol: acetic acid by volume.
16. 70% (v/v) ethyl alcohol.

2.2 Solutions for Mammalian Cell Culture

1. RPMI-1640 without glutamine (store at room temperature). Alternatively 10× RPMI without glutamine, sodium bicarbonate, or folic acid (store at 4 °C).
2. 7.5% sterile sodium bicarbonate ($NaHCO_3$): Store at room temperature.

3. Penicillin/streptomycin solution (10,000 U/mL penicillin, 10,000 μg/mL streptomycin): Store at −20 °C.
4. 200 mM glutamine: Store at −20 °C.
5. Heat-inactivated newborn calf serum (NCS) or fetal bovine serum (FBS): Store at −20 °C. To heat-inactivate, thaw the bottle of serum and place it in a water bath at 56 °C so that the water in the bath is above the level of the serum in the bottle. Incubate for 30 min, swirling every 5 min to ensure even heating.
6. 1 M Hepes (adjusted to pH 7.4 with NaOH): Filter sterilize, store at 4 °C.
7. 250 mg/L folic acid: Filter sterilize, store dark at 4 °C.

2.3 Media for HeLa S3 and Mouse FT210

1. Thaw frozen components in a 37 °C water bath.
2. For HeLa S3, add the following to 500 mL of 1× RPMI-1640 medium without glutamine: 5 mL penicillin/streptomycin solution, 5 mL 200 mM glutamine, and 25 mL serum (either NCS or FBS).
3. Alternatively, to 435.3 mL of sterile water (autoclaved in a 500 mL media bottle) add 50 mL 10× RPMI-1640 (without glutamine, $NaHCO_3$, or folic acid), 12.3 mL 7.5% sodium bicarbonate, 5 mL penicillin/streptomycin solution, 5 mL 200 mM glutamine, 25 mL serum (NCS or FBS), and 2.4 mL of 250 mg/L folic acid solution.
4. For FT210 cells, make 500 mL media the same as for HeLa cells except use 50 mL serum (NCS or FBS) and add 10 mL 1 M Hepes pH 7.4.
5. Store medium at 4 °C. Warm to 37 °C in incubator or water bath before adding to cells.
6. Freezer medium: Mix 20 mL serum plus 5 mL sterile DMSO. Store at −20 °C.

2.4 Media for Budding Yeast

1. Rich (YPAD) medium: Combine 5 g yeast extract, 10 g peptone, 50 mg adenine, and 10 g glucose (dextrose) in a 500 mL media bottle with 500 mL of water. Autoclave to sterilize, which will dissolve the components.
2. YPAD agar plates: Use the same components as for YPAD medium plus 17 g Bacto agar. After autoclaving to sterilize and dissolve the components, place the 500 mL bottle in a water bath at 56 °C until equilibrated. Pour into polystyrene petri dishes on a benchtop, allow the agar to set, and store the plates inverted at 4 °C.
3. CSM (Complete Supplementary Mix): Combine 2.0 g each of adenine hemisulfate, arginine HCl, aspartic acid, histidine HCl,

isoleucine, lysine HCl, methionine, serine, threonine, and tyrosine. Also add 3.0 g each of phenylalanine and tryptophan, 4.0 g leucine, 1.2 g uracil, and 9.0 g valine. Mix the components thoroughly with a spatula and grind with a mortar and pestle. Store at room temperature in a dessicator. (It may be necessary to omit one or more of histidine, leucine, lysine, tryptophan, and uracil. See **Note 1**).

4. 40% glucose solution: Heat 300 mL water in a microwave oven, add 200 g glucose, and then heat further and stir until mostly dissolved. Adjust the volume to 500 mL and autoclave.
5. Synthetic Complete (SC) medium: Combine 0.85 g yeast nitrogen base (YNB), 2.50 g ammonium sulfate, and 0.50 g CSM with 475 mL water in a 500 mL media bottle and autoclave. After it cools, add 25 mL 40% glucose. If histidine, leucine, lysine, tryptophan, and/or uracil have been omitted, add them as needed (*see* **Note 1**).
6. SC Petri plates: Use the recipe for SC medium above but add 17 g Bacto agar before autoclaving. Afterwards, place the bottle in a 56 °C water bath to equilibrate and pour plates as described in Subheading 2.4, **item 2**.

2.5 Equipment

1. Research microscope with 40× and 100× objectives, phase contrast, and epifluorescence. We use a Nikon Labophot 2.
2. Inverted microscope, to check cell cultures in T75 flasks.
3. Microscope slides and cover slips; hemacytometer and cover slips.
4. Humidified incubator providing 5% CO_2 and temperature control, for FT210 cultures and for starting HeLa S3 cultures.
5. For HeLa suspension cultures: an incubator, warm room, or water bath which can accommodate magnetic stirring platforms (5% CO_2 not needed).
6. For growth of yeast on Petri plates, an incubator at 28 °C.
7. Platinum loop; forceps.
8. Rotary water bath shaker.
9. Media bottles 250 mL and 500 mL, some sterilized with Teflon-coated magnetic stirring bars inside (1.5 and 2 inches, respectively).
10. Sterile 75 cm^2 (T75) polystyrene culture flasks.
11. Sterile 50 mL polypropylene conical centrifuge tubes with caps.
12. Benchtop centrifuge with swinging bucket rotor to spin 50 mL centrifuge tubes, capable of at least 1400× *g*.
13. Microcentrifuge and 1.5 mL conical tubes.

14. Pipettors and sterile tips.
15. Sterile cotton-plugged serological pipets (10 mL and 25 mL); 9-inch sterile cotton-plugged Pasteur pipets.
16. Sterile 2 mL cryotubes.

3 Methods

3.1 HeLa S3 Cell Culture

1. To start HeLa cells, thaw a cryotube of frozen cells by inserting the tube into a hole in a flat piece of Styrofoam and floating it in a 37 °C water bath for 5 min. In the safety cabinet, clean the outside of the cryotube with 70% ethanol. Transfer cells with a Pasteur pipet to a T75 flask containing 15 mL medium. Incubate at 37 °C with 5% CO_2.
2. Determine the cell concentration by transferring a small volume of the culture to a hemacytometer with a Pasteur pipet (*see* **Note 2**). Count the cells using a 20× phase contrast objective.
3. Dilute the cells daily to 2×10^5/mL or less. Use additional T75 flasks if needed to build up the cell population (*see* **Note 3**).
4. A spinner culture in a 500 mL bottle requires at least 5×10^7 cells (half as many in a 250 mL bottle) so that an adequate concentration of CO_2 will be maintained without an exogenous supply. When sufficient cells are obtained in T75 flasks, transfer them to a sterile 500 mL media bottle containing a 2-inch magnetic stir bar or a 250 mL media bottle containing a 1.5 inch stir bar. Stir the cells gently (60–90 rpm) on a stirring platform in a warm room, water bath, or incubator at 37 °C. In the warm room or incubator, place the culture bottle on a flat piece of Styrofoam (about 0.5 cm thick) to insulate it from the heat of the stirrer motor.
5. For spinner cultures, use the hemacytometer to determine the cell concentration daily. For best results, dilute the culture daily to 2×10^5 cells/mL. Maintain a culture volume of 200–400 mL in a 500 mL bottle or 100–200 mL in a 250 mL bottle (*see* **Note 3**).
6. To freeze cells for later use, spin 50–100 mL of cell culture in the benchtop centrifuge at room temperature at 180× *g* for 3 min. Resuspend the pellet to about 1×10^7 cells/mL in cryoprotective freezer medium and dispense into cryovials with 0.5–1.0 mL per vial. Place at −80 °C with Styrofoam insulation so that the sample cools and freezes gradually (*see* **Note 4**).

3.2 Mouse FT20 Cell Culture

1. Start cultures of FT210 cells [9] similarly as HeLa S3 cell cultures (Subheading 3.1) but incubate them at 32 °C (*see* **Note 5**).
2. Maintain FT210 cultures in T75 flasks at between 15 and 60 mL per flask and dilute them daily to 2–3 × 10^5 cells/mL.
3. Freeze FT210 cells using the same procedure as for HeLa (Subheading 3.1, **step 6**).

3.3 Budding Yeast (Saccharomyces Cerevisiae) Culture

1. To "wake up" frozen yeast, remove a vial of cells from the −80 °C freezer. Use a sterile 1000 μL pipettor tip to scrape a small amount of frozen medium from the surface. Gently streak this across the surface of a YPAD plate. Incubate inverted at 28 °C (*see* **Note 6**).
2. To start a liquid culture, heat the platinum loop to red hot in a Bunsen burner flame, jab it in the agar to cool, and then use it to pick a colony and transfer it to 10–20 mL medium in a sterile 50 mL polypropylene centrifuge tube. Incubate at 28 °C and 180 rpm in the rotary water bath shaker overnight.
3. To prepare a frozen stock of cells, grow cells 4 days at 28 °C in 15 mL YPAD. Pellet the cells (1400× *g*, 5 min), resuspend in 2 mL of 50% YPAD, 25% glycerol, and divide into sterile microfuge tubes or cryovials. Store at −80 °C.

3.4 Metaphase-Arrested HeLa Cells Treated with Cdk1 Inhibitors

1. For metaphase arrest, prepare a 400 mL spinner culture at 2–2.5 × 10^5 cells/mL in a sterile media bottle with a 2-inch stir bar and add 10 mL of 100 mM thymidine (final concentration 2.5 mM) (e.g., at 5 p.m., Day 1). Incubate at 37 °C with gentle stirring.
2. After 20 h (e.g., 1 p.m., Day 2), pour the culture into sterile 50 mL centrifuge tubes (or larger, if available) in the safety cabinet and pellet the cells at room temperature in a swinging bucket rotor, 180× *g* for 3 min (*see* **Note 7**). Decant the supernatant, resuspend the cells in a small amount of sterile 0.9% NaCl by flicking the tube, and pool the cell suspensions from multiple tubes into one. Adjust the volume to 50 mL with 0.9% NaCl and pellet the cells again. Resuspend the pellet in 250 mL fresh prewarmed medium and continue incubation.
3. After 4 h (e.g., 5 p.m., Day 2), add nocodazole to a final concentration of 0.25 μg/mL. Continue incubation.
4. At 16–20 h after removal of TdR, check the cell count and mitotic index.
5. To most easily determine the mitotic index, mix 200 μL cell culture and 200 μL of water containing 20 μg/mL Hoechst 33342 in a microfuge tube. Mix by flicking the tube and let it sit for 5 min to allow hypotonic cell swelling. Then add 40 μL

fixative and mix by flicking. View the cells by phase contrast or epifluorescence microscopy with a 40× objective and count the numbers of metaphase and interphase (nucleated) cells. (To make cell counting easier, increase the concentration of cells by pelleting them at low speed in the microcentrifuge, withdrawing most of the supernatant, and flicking to resuspend the fixed cells).

6. To inhibit Cdk1 and induce exit from mitosis, add roscovitine [10] to a final concentration of 200 μM for 1 h. Verify exit from metaphase-arrest (e.g., Fig. 1c) by observing nuclear reassembly (e.g., Fig. 1d). Nearly all cells reassemble nuclei within 1 h

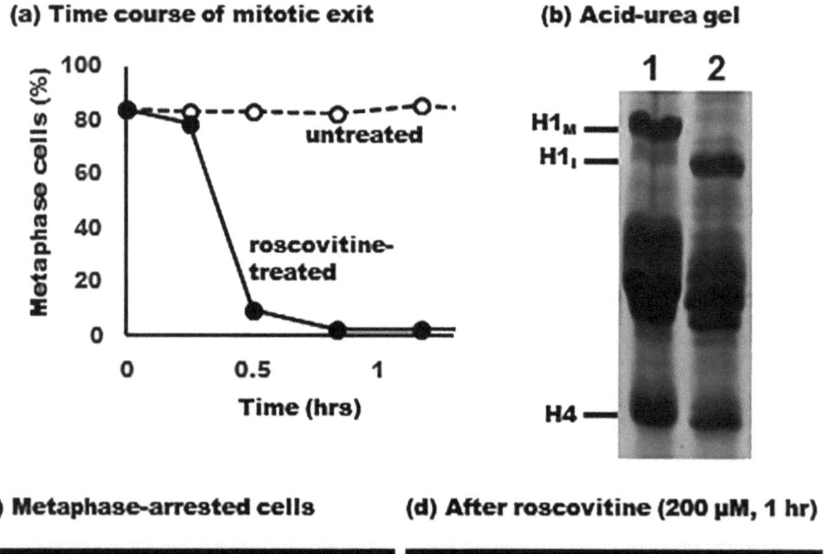

Fig. 1 Treatment with the Cdk1 inhibitor roscovitine induces metaphase-arrested HeLa S3 cells to exit mitosis. (**a**) Aliquots of a nocodazole-arrested culture (mitotic index, 85%) were either treated with 200 μM roscovitine (●—●) or incubated without treatment (○—○). Roscovitine-treated cells reassemble interphase nuclei within 1 h; (**b**) Histone H1 was phosphorylated in the metaphase-arrested cells (lane 1) but dephosphorylated in cells treated 1 h with roscovitine (lane 2); $H1_M$ indicates the position of phosphorylated H1 from mitotic cells. $H1_I$ indicates the position of interphase H1; (**c**) Nocodazole-arrested cells show condensed mitotic chromosomes; (**d**) Roscovitine treatment of nocodazole-arrested cells leads to reassembly of interphase nuclei. Cells were fixed and stained with Hoechst 33342 and viewed by epifluorescence microscopy

(Fig. 1a). Mitotic exit can also be confirmed by detection of histone dephosphorylation (Fig. 1b).

7. Alternatively, induce mitotic exit by adding staurosporine [11] to a final concentration of 500 nM [4], or by adding another inhibitor of Cdk1 (*see* **Note 8**).
8. Confirm mitotic exit using microscopy (see **Step 5** above) or other methods (*see* **Note 9**).

3.5 HeLa Cells Treated with Cdc25 Inhibitors

1. Cdc25 activity is required to keep Cdk1 active, and Cdc25 can be inhibited in vivo by sodium metavanadate [12] or menadione [13], leading to Cdk1 inactivation.
2. Arrest HeLa cells in metaphase as described in Subheading 3.4.
3. Add sodium orthovanadate to the culture and continue incubation, taking samples for mitotic index to follow the process of mitotic exit (e.g., Fig. 2a). Exit from mitosis is slower than with roscovitine and other Cdk1 inhibitors (Fig. 1 and Subheading 3.4). At least 2 mM orthovanadate is required for mitotic exit to be complete after 5 h (Fig. 2b, c).
4. Alternatively, treat with 100 μM menadione for 5 h or 200 μM for 3 h (Fig. 3).
5. Confirm exit from mitosis by microscopy or other methods (*see* **Note 9**).

3.6 Mouse FT210 Cells with Temperature-Sensitive Cdk1

1. For metaphase arrest, treat FT210 cultures in T75 flasks with 0.25 μg/mL nocodazole at 32 °C for 24–26 h. This typically yields cultures with mitotic indices of 70–90% with little or no loss of viability. Determine the mitotic index as described above for HeLa cells (Subheading 3.4, **step 5**).
2. To induce exit from mitosis, shift cells (in a stoppered Erlenmeyer flask) to high temperature between 40.0 and 42.0 °C [2] (Fig. 4). Incubate the cells in the rotary water bath shaker. The rate of mitotic exit depends on the temperature (see [2]) but is complete in 1.5 h at 40.6 °C (Fig. 4a).
3. Confirm exit from mitosis by microscopy (e.g., Fig. 4c, d), analysis of histone phosphorylation (e.g., Fig. 4b), or other methods (*see* **Note 9**).

3.7 Saccharomyces Cerevisiae with Analog-Sensitive Cdk1

1. Use a yeast strain carrying the allele *cdc28-as1*, which makes Cdc28 (Cdk1) sensitive to inhibition by the cell-permeable ATP analog 1NM-PP1 [7].
2. Dilute an overnight culture in SC, then grow to $OD_{600} = 0.4$. Treat the cells with 15 μg/mL nocodazole and continue incubation at 28 °C for 3 h.

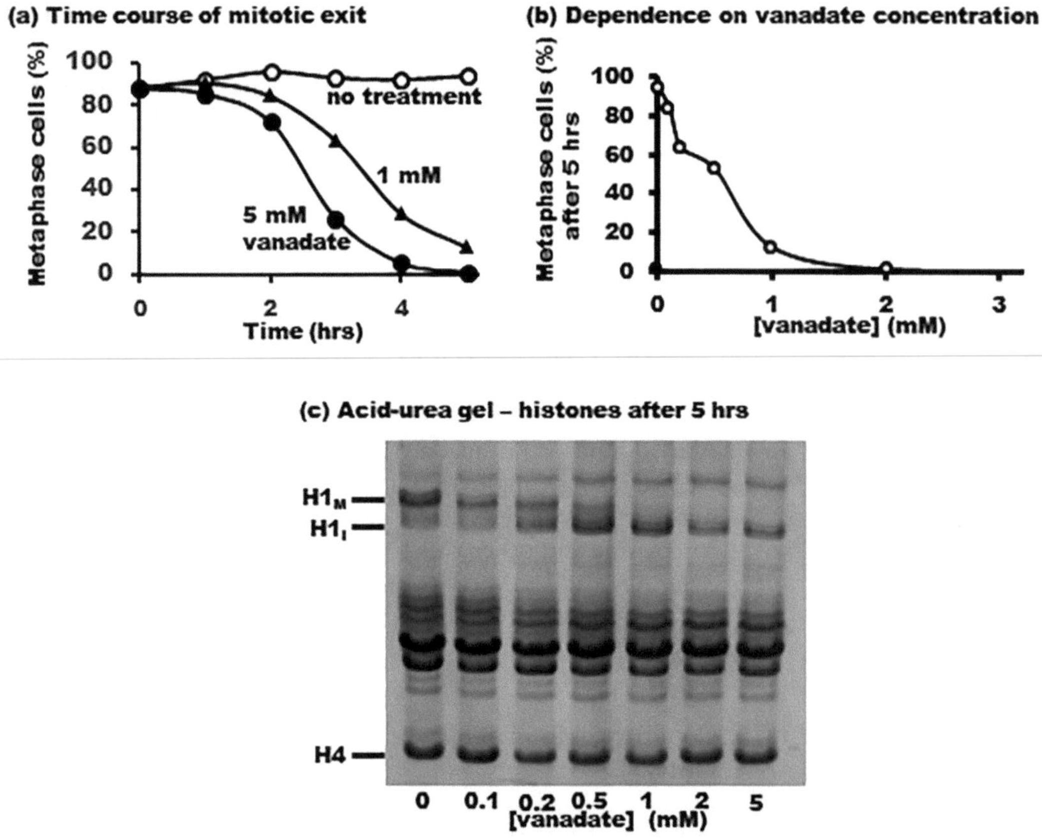

Fig. 2 Treatment of metaphase-arrested HeLa S3 cells with sodium orthovanadate, an inhibitor of the Cdc25 phosphatase, induces mitotic exit. (**a**) Percentage of metaphase-arrested cells as a function of time: No treatment (○—○),treatment with 5 mM vanadate (●—●), and treatment with 1 mM vanadate (▲—▲); (**b**) Percentage of cells remaining in metaphase arrest after treatment for 5 h with various concentrations of vanadate (○—○); (**c**) Acid-urea gel electrophoresis of acid-extracted histones showing the extent of histone H1 dephosphorylation after 5 hr. treatment with various concentrations of vanadate. $H1_M$ indicates the position of phosphorylated H1 from mitotic cells. $H1_I$ indicates the position of interphase H1

3. Check the percentage of cells in G2/M by mixing 100 μL of the culture with 20 μL of fixative [6]. G2/M cells appear as in Fig. 5b.
4. Treat the cells at 28 °C with at least 1.5 μM 1NM-PP1, either in a 50 mL capped polypropylene centrifuge tube or a stoppered Erlenmeyer flask. Incubate in the rotary water bath shaker at 180 rpm.
5. Confirm exit from mitosis by observing rebudding in the phase-contrast microscope (e.g., Fig. 5c) (*see* **Note 10**).
6. Alternatively, treat the cells with both 10 μM 1NM-PP1 and 100 μM α-factor and confirm mitotic exit by observation of mating projections ("shmoos") in the phase-contrast microscope (e.g., Fig. 5d) (*see* **Note 11**).

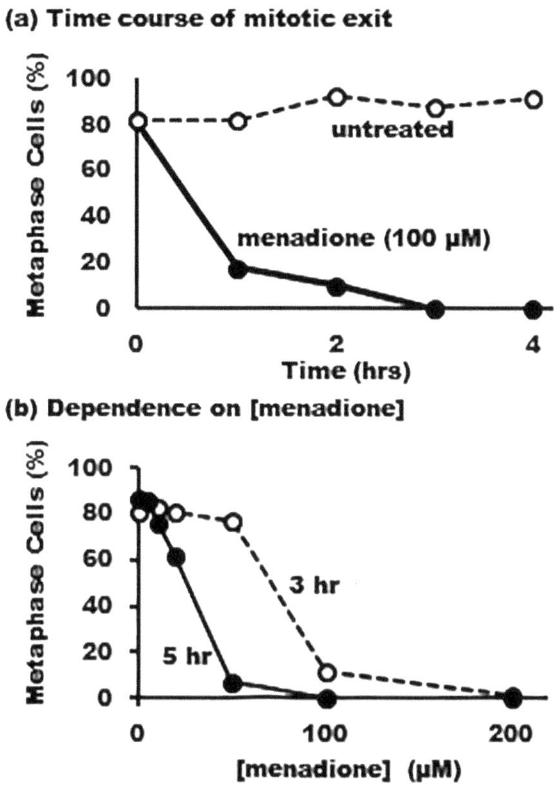

Fig. 3 The Cdc25 phosphatase inhibitor menadione induces mitotic exit and reassembly of nuclei in metaphase-arrested HeLa S3 cells. (**a**) Time course showing the percentage of cells in metaphase-arrest (condensed chromosomes, no nuclear envelopes): treatment with 100 μM menadione (●—●) or untreated (○—○). (**b**) Extent of exit from mitosis and reassembly of nuclei after 5 h (●—●) or 3 h (○—○) as a function of the concentration of menadione used in treatment

4 Notes

1. For yeast CSM, it may be important to omit one or more of histidine, leucine, tryptophan, lysine, and uracil for purposes of selection when doing crosses or transformations. If those components are omitted from CSM, then when making Synthetic Complete (SC) add them back as needed from separate stock solutions: 4 g/L histidine, 4 g/L leucine, 4 g/L lysine, 4 g/L tryptophan, and 2 g/L uracil. Those solutions should be autoclaved, except the tryptophan which should be filter sterilized. For 500 mL of SC, use 6.25 mL histidine solution, 12.5 mL leucine, 5.5 mL lysine, 9.375 mL tryptophan, and 7.5 mL uracil.

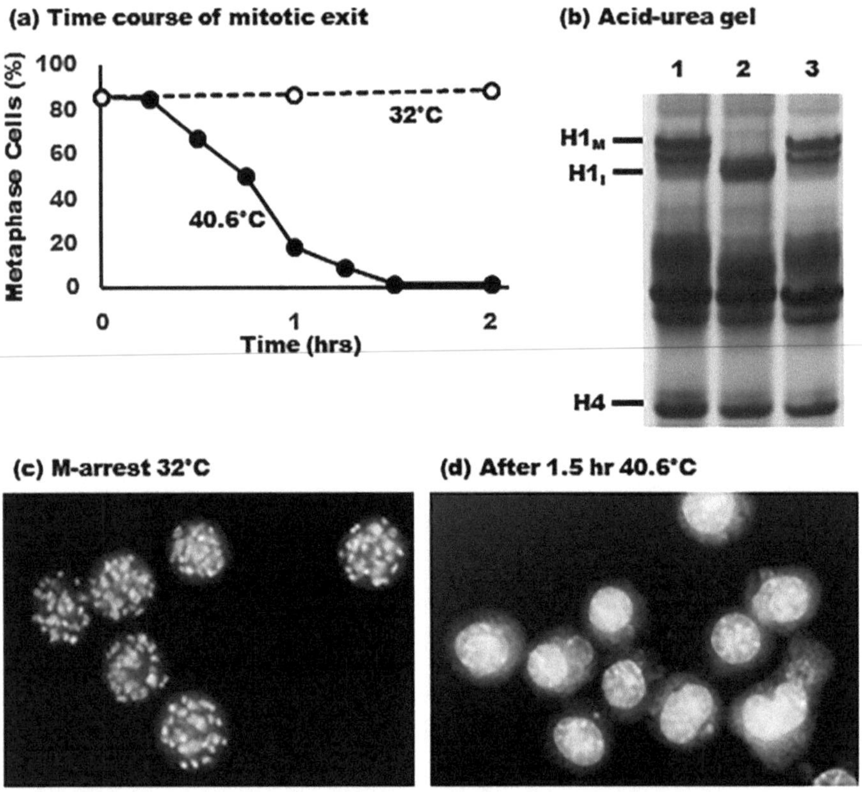

Fig. 4 Exit from mitosis induced by heat treatment of metaphase-arrested mouse FT210 cells, which have temperature sensitive Cdk1; (**a**) Time course showing the percentage of metaphase-arrested cells in culture aliquots incubated at 40.6 °C (●—●) and 32 °C (○—○); (**b**) Acid-urea gel electrophoresis of acid-extracted histones from metaphase-arrested cells at $T = 0$ (lane 1), cells incubated 1.5 h at 40.6 °C (lane 2), and cells as in lane 1 but incubated an additional 1.5 h at 32 °C (lane 3). $H1_M$ indicates the position of phosphorylated H1 from mitotic cells. $H1_I$ indicates the position of interphase H1; (**c**), (**d**) Micrographs of metaphase-arrested FT210 cells before (**c**) and after (**d**) heat treatment at 40.6 °C for 1.5 hr. (Adapted from [22])

2. Note that HeLa S3 cells grow in suspension. They attach weakly if at all to the surface in a T75 flask. Those that do attach can be dislodged into the liquid medium with a quick, vigorous shake before taking a sample for the hemacytometer.
3. Under these growth conditions, HeLa S3 cells double every 20–24 h. With suspension cultures not grown in a 5% CO_2 incubator, best growth is achieved if the cells are not diluted to less than 2×10^5 cells/mL and not allowed to exceed $6–8 \times 10^5$/mL (see Subheading 3.1, **step 5**). However, HeLa S3 cells grown in T75 flasks in a 5% CO_2 atmosphere can be diluted further, for example to 0.5×10^5/mL to be left over a weekend.
4. It has been found that the viability of mammalian cells is best preserved when they are frozen slowly (in cryoprotective

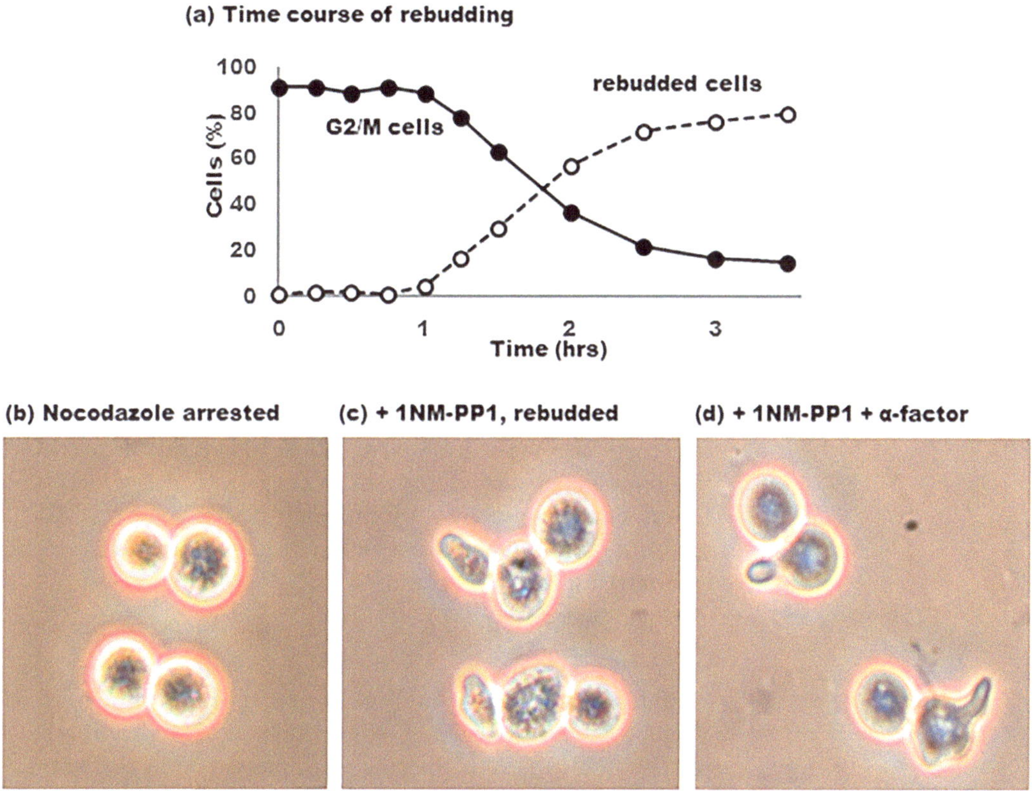

Fig. 5 Exit from mitosis following inhibition of Cdk1 in budding yeast (*S. cerevisiae*). The yeast strain used contained the allele *cdc28-as1* which makes Cdk1 (Cdc28 in *S. cerevisiae*) sensitive to inhibition by the ATP analog 1NM-PP1 ([7]). (**a**) Time course showing the decreasing percentage of nocodazole-arrested (G2/M) cells (●—●) and increasing percentage of rebudded cells (○—○). (Adapted from [6]). (**b**), (**c**), (**d**) Phase-contrast microscopy of metaphase-arrested (G2/M) cells (**b**), rebudded cells after 3 h treatment with 1.5 μM 1NM-PP1 (**c**), and "shmoos" (cells with mating projections) after 3 h treatment with both 1.5 μM 1NM-PP1 and 100 μM α-factor (**d**)

medium) and thawed quickly. The latter is achieved by floating a vial of frozen cells in a water bath at 37 °C for a few minutes (see Subheading 3.1, **step 1**). For cell freezing, gradual cooling at a rate of 1 °C per minute is considered best, and commercially available freezing containers ensure that rate. We use a simpler system consisting of two Corning Styrofoam racks for 50 mL centrifuge tubes placed face-to-face, horizontally, with the cryotubes upright in the wells of the rack. The two racks are taped together and placed in the −80 °C freezer with the cryotubes right-side up. After 24 h, the cells are frozen and the vials are transferred to freezer boxes at −80 °C for long-term storage. Cells are best stored at or near liquid N_2 temperature if that is available. At −80 °C significant loss of viability may occur after a year or more.

5. FT210 cells were originally reported by Mineo et al. [9]. They have a mutation which makes Cdk1 temperature sensitive. We obtained FT210 cells from P.M. Yau and E.M. Bradbury in the 1990s but later lost them due to a freezer failure. It is not clear where if at all they are available now, but other cells with ts-Cdk1 may substitute. On the other hand, inactivating Cdk1 via ts-mutation has a disadvantage. It would be incompatible with using cold- or heat-sensitive mutations to study other factors involved in mitotic exit.
6. Yeast strains are typically sent from lab to lab in the post, at ambient temperature. Using a sterile forceps, a few layers of sterile filter paper (pieces 1–2 cm square) are placed at the center of a piece of Saran or Cling wrap, at least 10 cm square, stretched flat. A few droplets from a liquid culture are added to the filter paper and the plastic wrap is folded over multiple times, using pressure to minimize air pockets. To start a culture, the plastic wrap is cut or unfolded, and the filter paper transferred to a YPAD plate.
7. The benchtop centrifuge does not need to be in the safety cabinet.
8. Several cell-permeable inhibitors of Cdk1 are known which could be used in place of roscovitine or staurosporine. For reviews see [14–16]. Specific examples include flavopiridols [17, 18] and purvalanol [19]. Vertebrate cell lines with analog-sensitive Cdk1 have also been made (e.g., [20, 21]) and could be used analogously as described for budding yeast in Subheading 3.7.
9. Mitotic exit for mammalian cells is most easily confirmed by microscopy but can also be done by analysis of histone phosphorylation (e.g., Figs. 1, 2, 4) or other approaches. When knocking out other factors (e.g., protein phosphatases) to see the effects, the same methods can be used to check for *failure* of mitotic exit.
10. Any concentration of 1NM-PP1 greater than 1.5 μM will induce metaphase-arrested *cdc28-as1* cells to exit mitosis. However, higher concentrations tend to repress budding. 1.5–2.0 μM 1NM-PP1 is optimal for observing rebudding [6].
11. Inducing "shmoos" with α-factor requires that the yeast be mating type a (MAT a).

References

1. Norbury C, Nurse P (1992) Animal cell cycles and their control. Annu Rev Biochem 61:441–470. https://doi.org/10.1146/annurev.bi.61.070192.002301
2. Paulson JR (2007) Inactivation of Cdk1/cyclin B in metaphase-arrested mouse FT210 cells induces exit from mitosis without chromosome segregation or cytokinesis and allows passage

through another cell cycle. Chromosoma 116: 215–225. https://doi.org/10.1007/s00412-006-0093-1

3. Shah JV, Cleveland DW (2000) Waiting for anaphase: Mad2 and the spindle assembly checkpoint. Cell 103:997–1000. https://doi.org/10.1016/s0092-8674(00)00202-6
4. Paulson JR, Patzlaff JS, Vallis AJ (1996) Evidence that the endogenous histone H1 phosphatase in HeLa mitotic chromosomes is protein phosphatase 1, not protein phosphatase 2A. J Cell Sci 109:1437–1447. https://doi.org/10.1242/jcs.109.6.1437
5. Wu JQ, Guo JY, Tang W, Yang CS, Freel CD, Chen C, Nairn AC, Kornbluth S (2009) PP1-mediated dephosphorylation of phosphoproteins at mitotic exit is controlled by inhibitor-1 and PP1 phosphorylation. Nat Cell Biol 11:644–651. https://doi.org/10.1038/ncb1871
6. Keaton JM, Workman B, Xie L, Paulson JR (2023) Analog-sensitive Cdk1 as a tool to study mitotic exit: protein phosphatase 1 is required downstream from Cdk1 inactivation in budding yeast. Chromosome Res 31:27–42. https://doi.org/10.1007/s10577-023-09736-6
7. Bishop AC, Ubersax UA, Petsch DT, Matheos DP, Gray NS, Blethrow J, Shimizu E, Tsien JZ, Schultz PG, Rose MD, Wood JL, Morgan DO, Shokat KM (2000) A chemical switch for inhibitor-sensitive alleles of any protein kinase. Nature 407:395–401. https://doi.org/10.1038/35030148
8. Touati SA, Kataria M, Jones AW, Snijders AP, Uhlmann F (2018) Phosphoproteome dynamics during mitotic exit in budding yeast. EMBO J 37:e98745. https://doi.org/10.15252/embj.201798745
9. Mineo C, Murakami Y, Ishimi Y, Hanaoka F, Yamada MA (1986) Isolation and analysis of a mammalian temperature-sensitive mutant defective in G2 functions. Exp Cell Res 167: 53–62. https://doi.org/10.1016/0014-4827(86)90203-x
10. Meijer L, Raymond E (2003) Roscovitine and other purines as kinase inhibitors. From starfish oocytes to clinical trials. Acc Chem Res 36: 417–425. https://doi.org/10.1021/ar0201198
11. Gadbois DM, Hamaguchi JR, Swank RA, Bradbury EM (1992) Staurosporine is a potent inhibitor of $p34^{cdc2}$ and $p34^{cdc2}$-like kinases. Biochem Biophys Res Commun 184:80–85. https://doi.org/10.1016/0006-291x(92)91160-r
12. Dunphy WG, Kumagai A (1991) The cdc25 protein contains an intrinsic phosphatase activity. Cell 67:189–196. https://doi.org/10.1016/0092-8674(91)90582-j
13. Wu FY, Sun TP (1999) Vitamin K3 induces cell cycle arrest and cell death by inhibiting Cdc25 phosphatase. Eur J Cancer 35:1388–1393. https://doi.org/10.1016/s0959-8049(99)00156-2
14. Meijer L (1995) Chemical inhibitors of cyclin-dependent kinases. Prog Cell Cycle Res 1:351–363. https://doi.org/10.1007/978-1-4615-1809-9_29
15. Hajdúch M, Havlíèek L, Veselý J, Novotný R, Mihál V, Strnad M (1999) Synthetic cyclin dependent kinase inhibitors. New generation of potent anti-cancer drugs. Adv Exp Med Biol 457:341–353
16. Roskoski R Jr (2019) Cyclin-dependent protein serine/threonine kinase inhibitors as anticancer drugs. Pharmacol Res 139:471–488. https://doi.org/10.1016/j.phrs.2018.11.035
17. Sedlacek H, Czech J, Naik R, Kaur G, Worland P, Losiewicz M, Parker B, Carlson B, Smith A, Senderowicz A, Sausville E (1996) Flavopiridol (L86 8275; NSC 649890), a new kinase inhibitor for tumor therapy. Int J Oncol 9:1143–1168. https://doi.org/10.3892/ijo.9.6.1143
18. Kim KS, Sack JS, Tokarski JS, Qian L, Chao ST, Leith L, Kelly YF, Misra RN, Hunt JT, Kimball SD, Humphreys WG, Wautlet BS, Mulheron JG, Webster KR (2000) Thio- and oxoflavopiridols, cyclin-dependent kinase 1-selective inhibitors: synthesis and biological effects. J Med Chem 43:4126–4134. https://doi.org/10.1021/jm000231g
19. Chen X, Liao Y, Long D, Yu T, Shen F, Lin X (2017) The Cdc2/Cdk1 inhibitor, purvalanol A, enhances the cytotoxic effects of taxol through Op18/stathmin in non-small cell lung cancer cells in vitro. Int J Mol Med 40:235–242. https://doi.org/10.3892/ijmm.2017.2989
20. de Castro IJ, Gil RS, Ligammari L, Di Giacinto ML, Vagnarelli P (2018) CDK1 and PLK1 coordinate the disassembly and reassembly of the nuclear envelope in vertebrate mitosis. Oncotarget 9:7763–7773. https://doi.org/10.18632/oncotarget.23666

21. Samejima I, Spanos C, Samejima K, Rappsilber J, Kustatscher G, Earnshaw WC (2022) Mapping the invisible chromatin transactions of prophase chromosome remodeling. Mol Cell 82:696–708. https://doi.org/10.1016/j.molcel.2021.12.039

22. Paulson JR, Vander Mause ER (2013) Calyculin A induces prematurely condensed chromosomes without histone H1 phosphorylation in mammalian G1-phase cells. Adv Biol Chem 3: 36–43. https://doi.org/10.4236/abc.2013.33A005

Chapter 16

Targeting Mitotic Exit in Malignant Cells

Christine Greil, Monika Engelhardt, Julia Felthaus, and Ralph Wäsch

Abstract

In order to sustain genomic stability by correct DNA replication and mitosis and thus avoid malignant transformation of cells, the cell cycle is a strictly regulated process. Aberrant cell cycle regulation and defects in mitosis in malignant cells are targets of various cancer therapies. Cancer cells may survive antimitotic treatment due to mitotic slippage with a residual activity of the ubiquitin ligase anaphase-promoting complex (APC/C) and a continuous slow ubiquitin-proteasome-dependent cyclin B-degradation leading to mitotic exit. The combination of antimitotic chemotherapeutics with proteasome inhibitors to block cyclin B-proteolysis or with targeted inhibitors of the APC/C and the antiapoptotic protein Mcl-1 seems a promising approach to improve treatment response in different malignancies by enhancing mitotic arrest and apoptosis.

The influence of conventional spindle poisons and new targeted substances and of their combinations on mitosis and apoptosis has not yet been conclusively clarified. Most models have been verified on cell lines whose biology may differ from that of tumors growing in vivo. To study the impact of various antimitotic substances on cell proliferation, especially detect onset of apoptosis depending on different cell cycle phases and thus to identify a possibly entity-dependent mechanism of those agents and their combinations, a combined approach with live-cell imaging and soft-agar colony assays in cultured patient-derived xenografts (PDX) was established.

Key words Mitotic exit, Antimitotic therapy, Mitotic slippage, Cyclin B, Anaphase-promoting complex/cyclosome (APC/C), Live-cell imaging, Soft-agar colony assay, Patient-derived xenograft (PDX)

1 Introduction

Targeting mitosis is one of the most common chemotherapeutic approaches in the treatment of malignant tumors. Several studies with cancer cell lines could prove a sustained mitotic arrest after treatment with high-dosed paclitaxel with subsequent cell death because of the known microtubule-stabilizing effect [1]. However, it was shown in breast cancer that in vivo attainable concentrations of paclitaxel do not induce a mitotic arrest, but multipolar spindles, resulting in chromosome missegregation and apoptosis during the subsequent interphase and that doses sufficient for a mitotic block are not attainable in vivo [2]. Nevertheless, there are potentially

Victor M. Bolanos-Garcia (ed.), *Mitotic Exit: Methods and Protocols*, Methods in Molecular Biology, vol. 2874,
https://doi.org/10.1007/978-1-0716-4236-8_16,

entity-dependent differences and at low doses spindle poisons may still prevent the spindle assembly checkpoint (SAC) from being satisfied due to disturbed microtubule kinetics with resulting defective chromosome attachment and free kinetochores [3] leading to the formation of the mitotic checkpoint complex (MCC). After binding to Mad2, BubR1, and Bub3 in the MCC, Cdc20 can no longer activate the anaphase-promoting complex/cyclosome (APC/C). APC/C inhibition leads to stabilization of securin and cyclin B, preventing chromosome segregation and mitotic exit. The resulting cell cycle delay can either trigger degradation of the antiapoptotic Bcl-2 family member Mcl-1, caspase activation and thus mitotic cell death [4] or result in mitotic exit despite incomplete cell division and proliferation of cells with a highly aberrant genome [5]. Moreover, SAC-deficient malignant cells may survive antimitotic treatment even more likely due to mitotic slippage with a residual APC/C activity and a continuous slow ubiquitin-proteasome-dependent cyclin B-degradation leading to mitotic exit [6]. Therefore, blocking cyclin B-proteolysis via additional proteasome or APC/C inhibition may prevent mitotic slippage, lead to a more robust mitotic block and enhance apoptosis in mitosis [7]. Additional inhibition of Mcl-1 may contribute to restore apoptosis in mitosis [8] (Fig. 1). To study entity-dependent effects of different substances targeting mitotic exit and their combinations on cell cycle and cell proliferation, a combined approach with live-cell imaging to differentiate between mitotic cell death and apoptosis during interphase (Fig. 2) and soft-agar colony assays to examine a possible long-term effect was established [7] in ex vivo cultures of patient-derived xenografts (PDX), as these are closest to the biology of in vivo growing tumors [9]. Therefore, the described experimental approach may improve in vitro drug testing to better predict successful in vivo cancer treatment.

2 Materials

2.1 Cell Culture

1. PDX from different entities: provided by Charles River Discovery Research Services Germany GmbH, Freiburg, Germany.
2. Growth medium: RPMI1640-medium with L-glutamine, 10% fetal bovine serum, 1% penicillin/streptomycin, 1 mM sodium pyruvate, and 25 mM Hepes. Optional: Sterile mineral oil (*see* **Note 1**).
3. Incubator set for cell culture conditions: 37 °C, 90% air humidity, 5% CO_2.
4. Microscope 8-well chambers with optical bottoms which have been tissue culture treated, for example, μ-Slide 8-well, Ibidi (Cat. 80826).

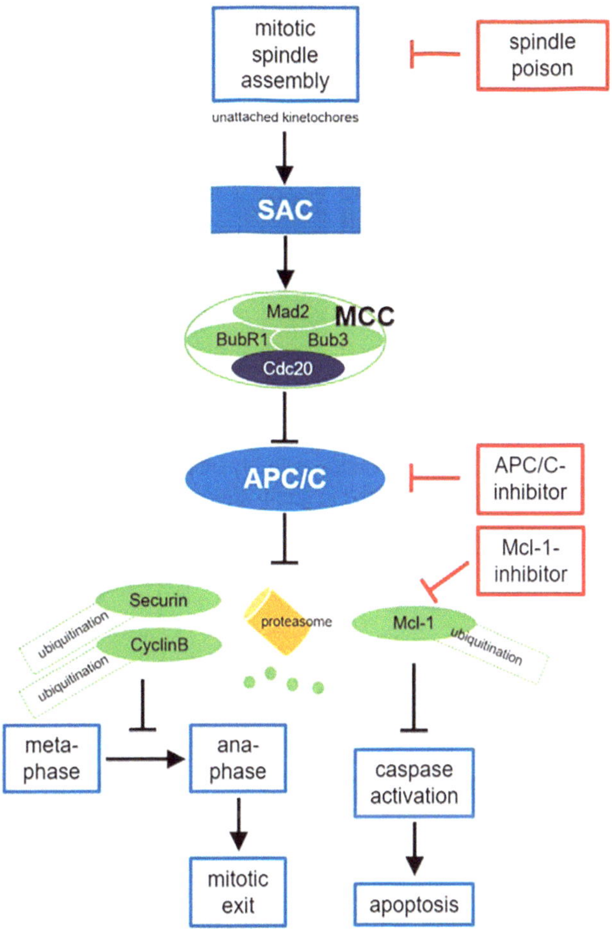

Fig. 1 Cancer treatment targeting mitotic exit. (Figure derived from Greil et al. [10]). Treatment with spindle poisons leads to disturbed microtubule kinetics and consequently to a defective chromosome attachment and free kinetochores. Thus, the SAC cannot be satisfied leading to MCC-formation consisting of Mad2, BubR1, Bub3, and Cdc20. Subsequently, Cdc20 can no longer activate the APC/C leading to securin and cyclin B stabilization and prevention of chromosome segregation and mitotic exit. Via Mcl-1-degradation, prolonged mitotic arrest leads to apoptosis in mitosis. Due to residual APC/C activity and slow cyclin B-degradation SAC-deficient malignant cells may survive treatment with spindle poisons. APC/C inhibition may prevent this mitotic slippage and thus enhance the antitumor effect. Apoptosis may be additionally promoted by Mcl-1 inhibition. Abbreviations: *SAC* spindle assembly checkpoint, *MCC* mitotic checkpoint complex, *APC/C* anaphase-promoting complex/cyclosome

2.2 Automated Microscope (Olympus ScanR) (See Note 2)

1. CY5-filter set (RFP-fluorescence).
2. SiR-DNA kit (*see* **Note 3**),
3. UPLSAPO objective × 20 (numerical aperture 0.75).

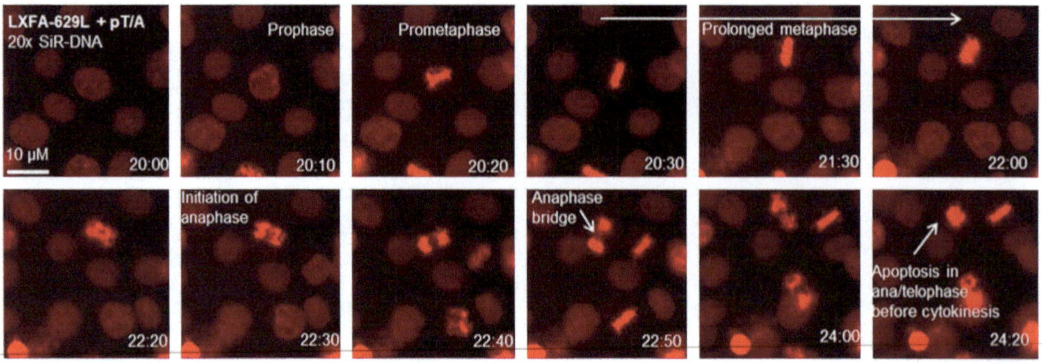

Fig. 2 The course of cell cycle leading to apoptosis is traced back on single cell level via live-cell imaging, here shown for the non-small cell lung cancer patient-derived xenograft LXFA-629 L after anaphase-promoting complex/cyclosome (APC/C) inhibition with proTAME/Apcin (pT/A). In this example, cell death is initiated in mitosis. The time point is indicated in h:min after addition of pT/A. (Figure derived from Greil et al. [7])

4. Microscope climate chamber to allow incubation of the samples at 37 °C and at least 60% humidity.
5. ScanR acquisition software.
6. ScanR analysis software.

2.3 Soft-Agar Colony Assay

1. 0.8% NuSieve™GTG™
2. 35 mm-culture dishes
3. 1.4% agarose.

2.4 Flow Cytometry

1. Propidium iodide containing 0.1 mg/ml RNaseA and 0.1% TritonX.
2. 70% ethanol.
3. BD LSRFortessa™ cell analyzer.
4. BD FACSDiva™ software.

3 Methods

3.1 Live Cell Imaging

1. Test PDX for mycoplasma and culture them in RPMI1640-medium with the appropriate additives under standard conditions. Trypsinize subconfluent cells which have been allowed to grow asynchronously for at least 48 h. To achieve an even distribution across the entire surface of the microscope chamber, centrifuge 30–50,000 cells at 300× *g* for 5 min, resuspend them in 350 μL growth medium, and transfer the cell suspension to an 8-well microscope chamber. Depending on the total

number of wells required for the experiment, adjust the cell number and the total volume of the suspension medium accordingly. Allow the seeded cells to grow at least 24 h in 8-well chambers under standard cell culture conditions.

2. Treat cells with the appropriate substances as indicated in Subheading 3.3.
3. Use the SiR-DNA kit for nucleus stain according to manufacturer's instruction.
4. Adjust air temperature of the climate chamber of an Olympus IX-81 inverse microscope at least 2 h prior to the planned experiment to 37 °C while still in dry mode. Preheating before setting the humidity is important to avoid condensation and subsequent damage to the microscope. After the microscope has reached 37 °C, adjust the air humidity to 60% and CO_2 level to 5%.
5. Start the ScanR acquisition software and define settings optimized for use with the UPLSAPO objective x 20. Define which wells are to be analyzed, the number of images per well and their spacing within the well, the acquisition cycle time, and the absolute number of acquisition cycles (*see* **Note 4**).
6. Start the image acquisition and supervise the first two cycles to ensure that the automatic focus is functioning well. The microscope should be set to focus on RFP-DNA signal using a CY5-filter set for all positions within a well and for each of the wells to be examined, before the next cycle starts.
7. Start ScanR analysis software (*see* **Note 4**).
8. Looking at larger numbers of cells allows a representative and objective first viewing of the cell population. Select apoptotic cells and trace back the course of cell cycle leading to apoptosis on single cell level. For this purpose, generate an exportable picture gallery to illustrate RFP-DNA for every single time point for a cell of interest. Determine if apoptosis is initiated during interphase or in mitosis (Fig. 2).

3.2 Soft-Agar Colony Assays

1. Prepare a bottom layer with 0.8% NuSieve™GTG™ agarose in medium in 35 mm-culture dishes.
2. Plate 1 ml medium containing 50,000 cells after the appropriate treatment (Subheading 3.3) and after a wash step to remove the substances mixed with 1 ml 1.4% agarose in medium as top layer, cover it with medium and incubate under standard cell culture conditions.
3. Count the number of colonies at day 21 to determine the long-term effect of different substances and their combinations on cell proliferation.

3.3 Drug Administration

1. To determine the appropriate dose, treat the PDX of interest with rising concentrations of the spindle poison and its potential combination partner (*see* **Note 5**) and assess cell cycle distribution by DNA staining with propidium iodide. Choose the lowest dose of for example paclitaxel leading to an increased G2/M-arrest and the lowest dose of for example the proteasome inhibitor bortezomib leading to an increased apoptotic rate in the PDX for further experiments.
2. DNA content is measured based on fluorescence intensity by flow cytometry: Fix cells in 70% ethanol at 4 °C overnight, wash in PBS, incubate in 50 μg/mL propidium iodide containing 0.1 mg/mL RNaseA and 0.1% TritonX for 30 min at 37 ° C, and analyze a standard of 10,000 cells per sample or data point on a BD LSRFortessa™ cell analyzer. Software analysis of cell cycle distribution is performed with BD FACSDiva™ software.
3. Dilute paclitaxel in medium to the appropriate dose and treat cells for 24 h. After a wash step with medium, add bortezomib diluted accordingly in the appropriate medium for another 24 h.

4 Notes

1. For imaging over time periods longer than 48 h a small layer of sterile mineral oil can be added to limit the amount of evaporation. The mineral oil should previously be shaken, mixed with normal culture media, and left overnight in order to saturate the oil with media components before using with the cells.
2. The system was established based on the ScanR software, but can be reproduced on any comparable microscope station.
3. The use of the SiR-DNA kit helps to ensure fast and easy autofocusing. Autofocus and intensity settings also depend on the type of cell and have to be determined separately for each PDX.
4. A good starting point to use for the cycle time is between 5 to 10 min. The shortest cycle time achievable will be limited by the total number of positions to be acquired, as well as by the exposure times needed for the images. For longer cell survival, exposure times should be minimized as much as possible.
5. The system was established with the spindle poison paclitaxel and the proteasome inhibitor bortezomib but can be adapted to other agents of interest with proposed impact on cell cycle and apoptosis.

References

1. Weaver BA (2007) How Taxol/paclitaxel kills cancer cells. Mol Biol Cell 25:2677–2681
2. Zasadil LM, Andersen KA, Yeum D, Rocque GB, Wilke LG, Tevaarwerk AJ, Raines RT, Burkard ME, Weaver BA (2014) Cytotoxicity of paclitaxel in breast cancer is due to chromosome missegregation on multipolar spindles. Sci Transl Med 6:229ra43
3. Stukenberg PT, Burke DJ (2015) Connecting the microtubule attachment status of each kinetochore to cell cycle arrest through the spindle assembly checkpoint. Chromosoma 124:463–480
4. Topham CH, Taylo SS (2013) Mitosis and apoptosis: how is the balance set? Curr Opin Cell Biol 25:780–785
5. Fujiwara T, Bandi M, Nitta M, Ivanova EV, Bronson RT, Pellman D (2005) Cytokinesis failure generating tetraploids promotes tumorigenesis in P53 -null cells. Nature 437:1043–1047
6. Gascoigne KE, Taylor SS (2008) Cancer cells display profound intra- and interline variation following prolonged exposure to antimitotic drugs. Cancer Cell 14:111–122
7. Greil C, Felthaus J, Follo M, Ihorst G, Ewerth D, Schüler J, Schnerch D, Duyster J, Engelhardt M, Wäsch R (2021) Targeting mitotic exit in solid tumors. Am J Cancer Res 11:3698–3710
8. Raab M, Kobayashi NF, Becker S, Kurunci-Csacsko E, Krämer A, Strebhard K, Sanhaji M (2020) Boosting the apoptotic response of high-grade serous ovarian cancers with CCNE1 amplification to paclitaxel in vitro by targeting APC/C and the pro-survival protein MCL-1. Int J Cancer 146:1086–1098
9. Lai Y, Wei X, Lin S, Qin L, Cheng L, Li P (2017) Current status and perspectives of patient-derived xenograft models in cancer research. J Hematol Oncol 10:106
10. Greil C, Engelhardt M, Wäsch R (2022) The role of the APC/C and its coactivators Cdh1 and Cdc20 in cancer development and therapy. Front Genet 13:941565

Chapter 17

The Use of APC/C Antagonists to Promote Mitotic Catastrophe in Cancer Cells

Ammar Mayah, Raúl Benito Arenas, Agatha Bastida, and Victor M. Bolanos-Garcia

Abstract

The multiprotein subunit E3 ubiquitin ligase Anaphase-Promoting Complex/Cyclosome (APC/C) plays a key role in the control of mitosis progression. APC/C is the ultimate effector of the Spindle Assembly Checkpoint (SAC), the signaling system of higher organisms including the human that monitors the proper attachment of chromosomes to microtubules during cell division. Defects in this process result in genome instability, aneuploidy, premature aging, and cancer. APC/C roles in the SAC require its activation by the protein Cdc20. Interfering with APC/C activation by Cdc20 impairs APC/C substrate recognition, resulting in a delayed mitotic exit and eventually inducing cell death. This may be advantageous for the treatment of cancer and malignancies associated with SAC dysregulation. Here we describe a protocol to interfere with mitotic exit through the use of commercially available (Apcin, proTAME) as well as innovative small molecules we have developed that function as antagonists of APC/C activation by Cdc20. We show that the use of these molecules alone and in combination is effective to promote mitotic catastrophe and suppress cell expansion in 2D and 3D (spheroids) cancer cells of different tissue origin, including breast, cervical, and ovarian cancer.

Key words Mitotic exit, Anaphase-Promoting Complex/Cyclosome (APC/C), Spindle Assembly Checkpoint (SAC), Cdc20, 3D cancer cells (spheroids), Apcin analogues

1 Introduction

The Anaphase-Promoting Complex/Cyclosome (APC/C) is a large multi-subunit E3 ubiquitin ligase complex that controls cell cycle progression by promoting the ubiquitination of specific substrates that are subsequently degraded by the proteasome [1, 2]. Tight regulation of this fundamental biological process requires the physical interaction of the APC/C with Cdc20, which acts as an activator of APC/C catalytic function (Fig. 1a–c). Formation of the APC/C-Cdc20 complex controls the transition from metaphase to anaphase and mitotic exit [3]. Interestingly, Cdc20 can also act as an inhibitor of the APC/C in response to

Victor M. Bolanos-Garcia (ed.), *Mitotic Exit: Methods and Protocols*, Methods in Molecular Biology, vol. 2874,
https://doi.org/10.1007/978-1-0716-4236-8_17,

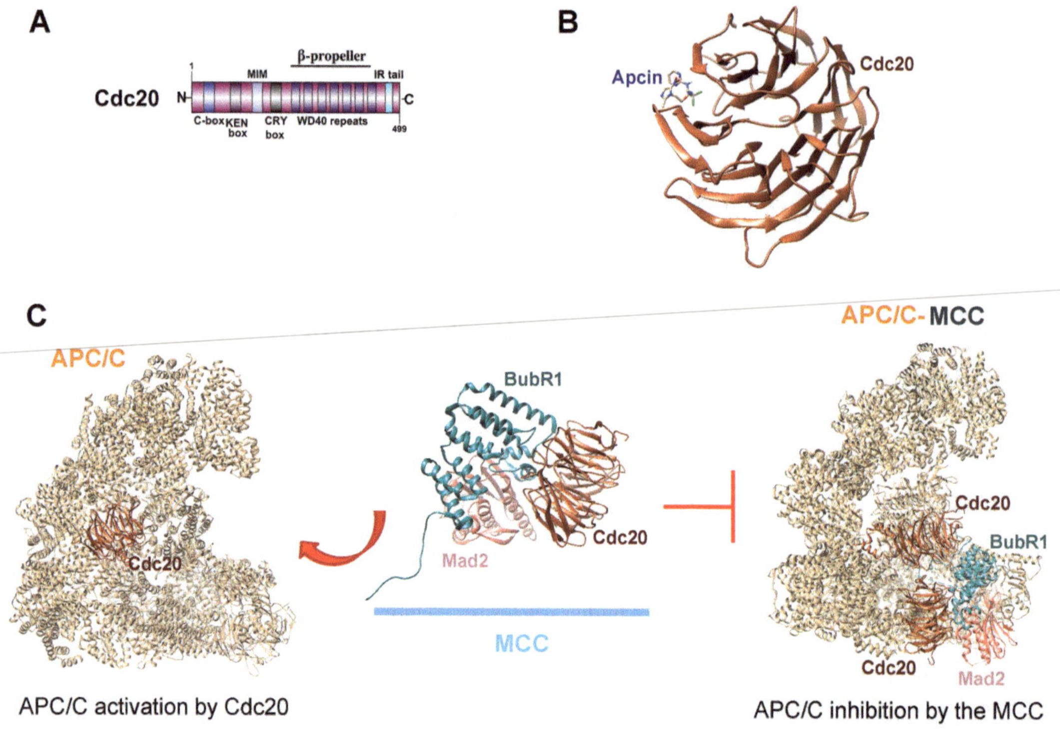

Fig. 1 APC/C-Cdc20 interactions. (**a**) Overall human Cdc20 domain organization. (**b**) Crystal structure of the Cdc20-Apcin complex (Protein Data Bank ID 4 N14). (**c**) 3D structures of APC/C alone and in complex with the Mitotic Checkpoint Complex (MCC) have provided valuable insights into APC/C regulation by Cdc20 (an interaction that results in APC/C activation) and by the MCC (an interaction that results in APC/C inhibition)

defects in chromosome attachment to the mitotic spindle, in a process that involves Cdc20 binding to Mad2, BubR1, and Bub3 to form the Mitotic Checkpoint Complex (MCC) [4, 5] (Fig. 1c). The direct interaction of the latter assembly prevents APC/C recognition and ubiquitination of its substrates, such as CyclinB1 and Securin, thus delaying anaphase onset [6].

Several small-sized compounds have been reported to inhibit APC/C functions through interference of APC/C activation by Cdc20 (Fig. 2) [7–10]. One of such compounds, TAME (tosyl-L-arginine methyl ester), acts as a mimetic drug of the Cdc20 C-terminal IR, which is implicated in Cdc20 binding to the APC/C, leading to inhibition of APC/C substrates degradation [11]. TAME has the great disadvantage of being poorly permeable to the cell membrane, a feature that stimulated to development of a prodrug known as proTAME. This prodrug can be processed in vivo by esterases to produce TAME inside the cell.

Apcin (2-(2-Methyl- 5-nitroimidazol-1-yl) ethyl N-[2,2,2-trichloro-1-(pyrimidin-2-ylamino) ethyl]carbamate) is another small-sized synthetic compound that directly affects APC/C E3 ubiquitin ligase activity. Apcin mechanism of action involves its binding to

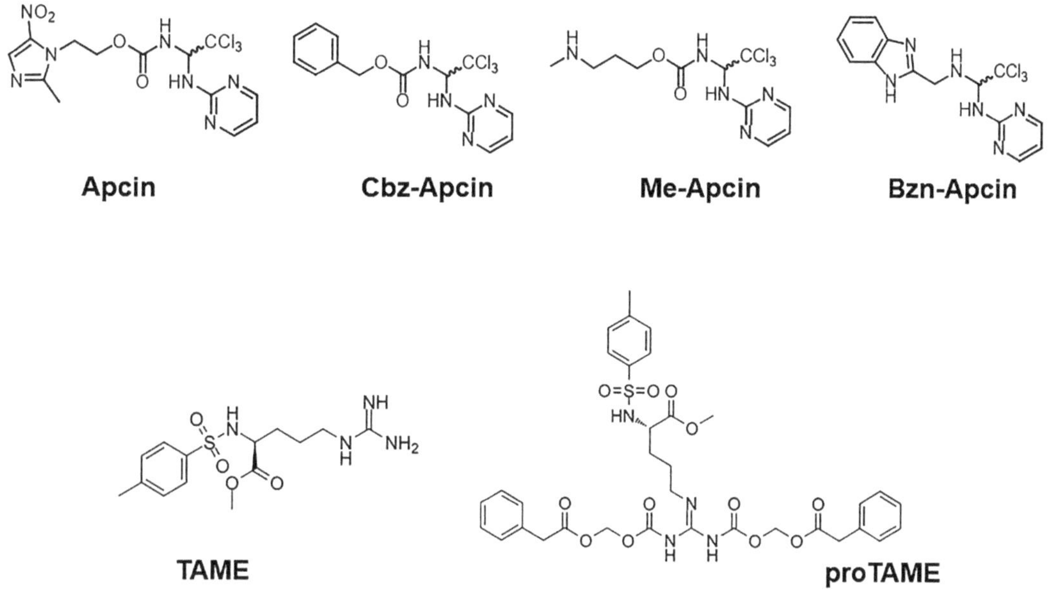

Fig. 2 Chemical structures of APC/C antagonists. Apcin and the new analogues share structural features important for binding a Cdc20 degron box (D-box). TAME acts as a mimetic drug of the Cdc20 C-terminal IR motif, which is implicated in Cdc20 binding to the APC/C. The very low cell membrane permeability of this molecule is overcome in the prodrug proTAME, a compound that is processed in vivo by esterases to produce active TAME inside the cell

a D-box degron motif of Cdc20 [12], an interaction that blocks substrate-induced Cdc20 loading onto the APC/C. The occupancy of the D-box degron leads to the inhibition of ubiquitination of Cdc20 substrates and an extended time in late mitosis. The latter condition eventually induces apoptosis, which can be advantageous to promote mitotic catastrophe in cancer cells. In this protocol, we report that the individual use of Apcin and the new Apcin analogues we have synthesized restrain [13] cell proliferation in 2D and 3D (spheroids) of breast, cervical, and ovarian cancer cells (Figs. 2 and 3, respectively). We also report that the use of Apcin and the new Apcin analogues when used in combination with proTAME led to a synergistic cytotoxic effect in all the cancer cell types tested.

2 Materials

2.1 Cell Culture

1. Growth medium. For SKOV-3 cells: RPMI1640-medium with L-glutamine, 10% fetal bovine serum, 1% penicillin/streptomycin, 1 mM sodium pyruvate and 25 mM Hepes. For HeLa and MDA-MB-231 cells: Dulbecco's modified Eagle's medium (DMEM) with L-glutamine, 10% fetal bovine serum, 1% penicillin/streptomycin.

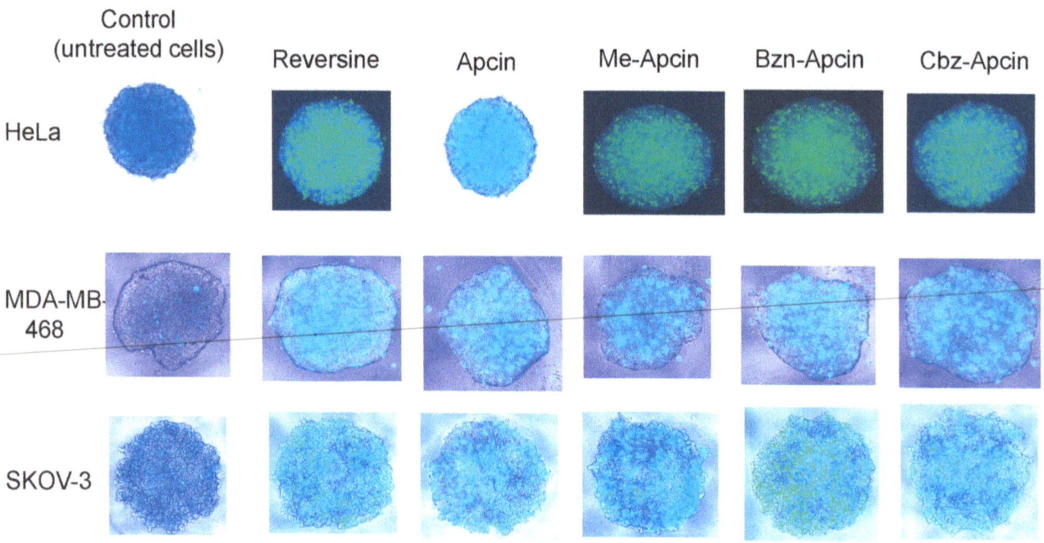

Fig. 3 Cytotoxicity of Apcin and new APC/C antagonists on cancer cell spheroids. 3D HeLa, MDA-MB-468, and SKOV-3 cancer cell spheroids. Reversine was used at 2 μM concentration. The rest of the molecules at 25 μM

2. Incubator set for cell culture conditions: 37 °C, 5% CO_2, 95% relative humidity.
3. Revolve microscope.

2.2 Cytotoxicity Assay

1. 3-(4,5-dimethylthiazol-2-yl)-2,5-diphenyl-2H-tetrazolium bromide (MTT).
2. 96-well plates.

2.3 Flow Cytometry

1. Propidium iodide.
2. Annexin V-FITC Apoptosis kit.
3. 4% of formaldehyde.
4. Hoechst 33258.
5. FxCycle PI/RNase staining kit.
6. NovoCyte Flow Cytometer.
7. FlowlogicTM Analysis Software.

2.4 Spheroids

1. Sytox Green.
2. U-shaped bottom designed for 3D culture.

2.5 Preparation of Compounds

All compounds solutions are prepared at room temperature using ultrapure DMSO. All the reagents required are of analytical grade quality.

1. Weight in an analytical balance the required amount to prepare 1 mL of compound stock solution at 10 mM concentration.

2. Resuspend each compound in DMSO and sterilize the resuspension by exposure to UV radiation.
3. Sonicate for 2 min in a sonicator.
4. In a tissue culture hood, prepare 700 uL of working solutions of the compounds at these concentrations: 50uM for Apcin and analogues; 10 uM for proTAME, using the appropriate mammalian cell growth medium. Vortex all the working solutions thoroughly.
5. Using a multichannel pipette, remove the entire medium (100 uL) from the desired wells of a 96-well plate previously seeded with the desired cancer cell line, making sure the confluence is not higher than 89%.
6. Add 100 uL of the compound working solution to each well. Typically, the 6 wells of one column are used to test one compound.
7. Incubate the plate for an additional 72 h.

2.6 MTT Assay

1. 3 h before end of treatment exposure, take the plates from the incubator (one at the time) and add 5 μL of MTT to each well.
2. Place the plates back in the incubator for the remaining treatment exposure time (e.g., 3 h).
3. Take the plates from the incubator and aspire the solution from each well using a multichannel pipette.
4. Add 100 μL of DMSO to each well.
5. Placed the plates on a shaker incubator for 15 min, room temperature.
6. Once a homogenous color is visible for each well, place the plate in a plate reader and record the absorbance at 570 nm.
7. Calculated the percentage of cell viability using the following equation:

$$\text{Viability of cells }(\%) = [(\text{Fluorescence of sample} - \text{Blank})/(\text{Fluorescence of negative control} - \text{Blank})] \times 100$$

2.7 Experiments in Spheroids

Day 1: Cell Seeding and Initial Incubation

1. Seed the cells at a density of 7×10^3 cells per well in a 100 μL volume in a 96-well plate with a U-shaped bottom designed for 3D culture compatibility.
2. Centrifuge the plate at 290× *g* for 3 min.
3. Incubate the plate for 24 h.

Day 2: Cell Treatment and Staining

1. Treat the cells with your desired experimental treatment.
2. Prepare a 250 nM Sytox Green dye solution by adding 0.5 μL of Sytox Green to 1 mL of cell culture media.
3. Add 20 μL of the prepared Sytox Green solution to each well containing 100 μL of media or 40 μL to each well containing 200 μL of media.
4. Centrifuge the plate at 250× *g* for 3 min.
5. Incubate the plate for an additional 24 h.

Note 1: Optional Collagen Gel Addition

If using collagen gel, prepare a 6 μg/mL collagen solution.

Add 100 μL of the prepared collagen solution to each well containing 100 mL of media to achieve a final collagen concentration of 3 μg/mL in each well (each well should now have a total volume of 200 μL).

Note 2: Treatment Preparation with Collagen Gel

When using collagen gel, prepare the treatment using media containing 6 μg/mL of collagen.

Day 3: Hoechst Staining

1. Prepare a stock solution of Hoechst by dissolving 1 mg of Hoechst dye in 1 mL of phosphate-buffered saline (PBS).
2. Create a working solution by adding 2 μL of the stock Hoechst solution to 1 mL of complete growth media, resulting in a final Hoechst concentration of 2 μg/mL.
3. Add 10 μL of the working Hoechst solution to each well containing 100 μL of media or 20 μL to each well containing 200 μL of media.
4. Centrifuge the plate at 250× *g* for 3 min.
5. Incubate the plate for 3–4 h.
6. After incubation, examine the cells under a fluorescent microscope or perform flow cytometry analysis or/and microplate reader.

Fluorescent Microscopy Use a 365 nm fluorescent filter for Hoechst and FITC (488–514 nm).

3 Notes

1. Optional Collagen Gel Addition

   If using collagen gel, prepare a 6 μg/mL collagen solution. Add 100 μL of the prepared collagen solution to each well containing 100 mL of media to achieve a final collagen concentration of 3 μg/mL in each well (each well should now have a total volume of 200 μL) [14].

2. Treatment Preparation with Collagen Gel

   When using collagen gel, prepare your treatment using media containing 6 μg/mL of collagen [14].

References

1. Peters JM (2006) The anaphase promoting complex/cyclosome: a machine designed to destroy. Nat Rev Mol Cell Biol 7:644–656
2. Barford D (2020) Structural interconversions of the anaphase-promoting complex/cyclosome (APC/C) regulate cell cycle transitions. Curr Opin Struct Biol 61:86–97
3. Alfieri C, Chang L, Zhang Z, Yang J, Maslen S, Skehel M, Barford D (2016) Molecular basis of APC/C regulation by the spindle assembly checkpoint. Nature 536:431–436
4. Lara-Gonzalez P, Pines J, Desai A (2021) Spindle assembly checkpoint activation and silencing at kinetochores. Semin Cell Dev Biol 117: 86–98
5. Kapanidou M, Curtis NL, Bolanos-Garcia VM (2017) Cdc20: at the crossroads between chromosome segregation and mitotic exit. Trends Biochem Sci 42:193–205
6. Curtis NL, Bolanos-Garcia VM (2019) The anaphase promoting complex/Cyclosome (APC/C): a versatile E3 ubiquitin ligase. Subcell Biochem 93:539–623
7. Wang L, Zhang J, Wan L, Zhou X, Wang Z, Wei W (2015) Targeting Cdc20 as a novel cancer therapeutic strategy. Pharmacol Ther 151:141–151
8. Bhuniya R, Yuan X, Bai L, Howie KL, Wang R, Li W, Park F, Yang CY (2022) Design, synthesis, and biological evaluation of Apcin-based CDC20 inhibitors. ACS Med Chem Lett 13: 188–195
9. Huang P, Le X, Huang F, Yang J, Yang H, Ma J, Hu G, Li Q, Chen Z (2020) Discovery of a dual tubulin polymerization and cell division cycle 20 homologue inhibitor via structural modification on Apcin. J Med Chem 63: 4685–4700
10. He Y, Le X, Hu G, Li Q, Chen Z (2023) Discovery of Ureido-based Apcin analogues as Cdc20-specific inhibitors against cancer. Pharmaceuticals (Basel) 16:304
11. Zeng X, Sigoillot F, Gaur S, Choi SW, Pfaff KL, Oh DC, Hathaway NA, Nevena D, Cuny GD, King RW (2010) Pharmacologic inhibition of the anaphase-promoting complex induces a spindle checkpoint-dependent mitotic arrest in the absence of spindle damage. Cancer Cell 18:382–395
12. Sackton KL, Dimova N, Zeng X, Tian W, Zhang M, Sackton TB, Meaders J, Pfaff KL, Sigoillot F, Yu H, Luo X, King RW (2014) Synergistic blockade of mitotic exit by two chemical inhibitors of the APC/C. Nature 514:646–649
13. Kapanidou M, Curtis LC, Diaz-Minguez SS, Agudo-Alvarez S, Rus Sanchez A, Mayah A, Agena R, Brennan P, Morales P, Benito-Arenas R, Bastida A, Bolanos-Garcia VM (2024) Targeting APC/C Ubiquitin E3-Ligase Activation with Pyrimidinethylcarbamate Apcin Analogues for the Treatment of Breast Cancer. Biomol 14:11
14. Nazari SS (2020) Generation of 3D Tumor spheroids with encapsulating basement membranes for invasion studies. Curr Protoc Cell Biol 87:e105

INDEX

Victor M. Bolanos-Garcia (ed.), *Mitotic Exit: Methods and Protocols*, Methods in Molecular Biology, vol. 2874,
https://doi.org/10.1007/978-1-0716-4236-8,

MIX
Papier aus verantwortungsvollen Quellen
Paper from responsible sources
FSC® C105338

If you have any concerns about our products,
you can contact us on
ProductSafety@springernature.com

In case Publisher is established outside the EU,
the EU authorized representative is:
Springer Nature Customer Service Center GmbH
Europaplatz 3, 69115 Heidelberg, Germany

Printed by Libri Plureos GmbH
in Hamburg, Germany